Springer Tracts in Electrical and Electronics Engineering

Series Editors

Springer Tracts in Electrical and Electronics Engineering (STEEE) publishes the latest developments in Electrical and Electronics Engineering - quickly, informally and with high quality. The intent is to cover all the main branches of electrical and electronics engineering, both theoretical and applied, including:

- Signal, Speech and Image Processing
- Speech and Audio Processing
- Image Processing
- Human-Machine Interfaces
- Digital and Analog Signal Processing
- Microwaves, RF Engineering and Optical Communications
- Electronics and Microelectronics, Instrumentation
- Electronic Circuits and Systems
- Embedded Systems
- Electronics Design and Verification
- Cyber-Physical Systems
- Electrical Power Engineering
- Power Electronics
- Photovoltaics
- Energy Grids and Networks
- Electrical Machines
- Control, Robotics, Automation
- Robotic Engineering
- Mechatronics
- Control and Systems Theory
- Automation
- Communications Engineering, Networks
- Wireless and Mobile Communication
- Internet of Things
- Computer Networks

Within the scope of the series are monographs, professional books or graduate textbooks, edited volumes as well as outstanding PhD theses and books purposely devoted to support education in electrical and electronics engineering at graduate and post-graduate levels.

Review Process

The proposal for each volume is reviewed by the main editor and/or the advisory board. The books of this series are reviewed in a single blind peer review process.

Ethics Statement for this series can be found in the Springer standard guidelines here https://www.springer.com/us/authors-editors/journal-author/journal-author-helpdesk/before-you-start/before-you-start/1330#c14214

Puvvula Vidyasagar · K. Shanti Swarup

Design and Development of Model Predictive Primary Control of Micro Grids

Simulation Examples in MATLAB

Springer

Puvvula Vidyasagar
Department of Electrical Engineering
Indian Institute of Technology Madras
Chennai, Tamil Nadu, India

K. Shanti Swarup
Department of Electrical Engineering
Indian Institute of Technology Madras
Chennai, Tamil Nadu, India

ISSN 2731-4200 ISSN 2731-4219 (electronic)
Springer Tracts in Electrical and Electronics Engineering
ISBN 978-981-19-5854-0 ISBN 978-981-19-5852-6 (eBook)
https://doi.org/10.1007/978-981-19-5852-6

This Springer imprint is published by the registered company Springer Nature Singapore Pte Ltd.
The registered company address is: 152 Beach Road, #21-01/04 Gateway East, Singapore 189721, Singapore

Contents

About the Authors

Puvvula Vidyasagar completed his Ph.D. from Indian Institute of Technology (IIT) Madras. He completed his M.Tech. from National Institute of Technology (NIT) Calicut and B.Tech. degree from RVR&JC Engineering College, India. His research interests are power systems analysis and control, smart grids, renewable energy technologies, micro grids, and power systems modelling. He has several research papers in journals and conferences published to his credit.

K. Shanti Swarup is a faculty with the Department of Electrical Engineering, Indian Institute of Technology (IIT) Madras, India. Before joining the department as a visiting faculty member, he held positions at the Mitsubishi Electric Corporation, Osaka, Japan, and Kitami Institute of Technology, Hokkaido, Japan, serving as a visiting research scientist and visiting professor, respectively, from 1992 to 1999. Since 2000, he has been a professor at IIT Madras. His research areas include power systems, smart grids, artificial intelligence, knowledge-based systems, computational intelligence, soft computing, Energy Management Systems (EMS), Supervisory Control and Data Acquisition (SCADA), power system automation, and network protection. He has done research projects with various industries like BHEL, Hitachi, Easun-MR, etc.

Abbreviations

AC	Alternating Current
DC	Direct Current
DER	Distributed Energy Resource
DG	Distributed generator
DMC	Dynamic Matrix Control
FIR	Finite Impulse Response
GPC	Generalized Predictive Control
IM	Induction Motor
kV_b	Base kV
L-MPC	Linear model-based MPC
LQR	Linear Quadratic Regulator
LTI	Linear Time-Invariant
LTI-MPC	Linear Time-Invariant Model Predictive Controller
LTV	Linear Time-Variant
LTV-MPC	Linear Time-Variant Model Predictive Controller
MPC	Model Predictive Controller
MPPT	Maximum Power Point Tracking
MVA_b	Base MVA
N-MPC	Nonlinear model-based MPC
OAT	One At a Time
PCC	Point of Common Coupling
P-f	Active power versus Frequency
PI	Proportional Integral
PID	Proportional Integral Derivative
PLL	Phase Locked Loop
PR	Proportional Resonant
PV	Photovoltaic
PV-DG	Photovoltaic Distributed Generator
PWM	Pulse Width Modulator
Q-V	Reactive power versus Voltage
R-L	Impedance load

SG-DG	Synchronous Generator based Distributed Generation
SHE	Selective Harmonic Elimination
VSC	Voltage Source Converter
VSI	Voltage Source Inverter
ZOH	Zero-Order Hold

Nomenclature

B1	Bus number 1
B2	Bus number 2
B3	Bus number 3
B4	Bus number 4
B5	Bus number 5
B6	Bus number 6
B7	Bus number 7
B8	Bus number 8
L2	Impedance load at bus B2
L5	Impedance load at bus B5
L6	Induction motor load at bus B6
L7	Impedance load at bus B7
L8	Impedance load at bus B8
ω_b	Angular speed of the global reference frame *D-Q* (rad/s)
$\boldsymbol{V}^{G1}$	Voltage space vector of the bus B1
ω_r	Angular speed of the SG-DG rotor reference frame $d1$-$q1$ (rad/s)
ω_3	Angular speed of the PV-DG local reference frame $d3$-$q3$ (rad/s)
$\boldsymbol{V}^{G3}$	Voltage space vector of the bus B3
δ_0	Torque angle or power angle of the SG-DG (rad)
δ_1	Rotor reference frame angle of the SG-DG (rad)
δ_3	Local reference frame angle of the PV-DG (rad)
f	Any arbitrary quantity (either voltage or current)
$\boldsymbol{f}_{abc}$	Vector of instantaneous values of the quantity f in stationary *abc* reference frame
$\boldsymbol{f}_{dq0}$	Vector of instantaneous values of the quantity f in rotating *d-q* reference frame
$\boldsymbol{T}_{abc\text{-}dq0}$	Park's transformation matrix
θ_c	Angle between *d-q* reference frame and *abc* reference frame
I_{d1}	$d1$-axis component of the SG-DG output current
I_{q1}	$q1$-axis component of the SG-DG output current
V_{d1}	$d1$-axis component of the bus B1 voltage

V_{q1}	$q1$-axis component of the bus B1 voltage
E'_{d1}	$d1$-axis component of the SG-DG stator induced voltage
E'_{q1}	$q1$-axis component of the SG-DG stator induced voltage
Ψ_{ad1}	$d1$-axis component of the damper winding flux linkage
Ψ_{aq2}	$q1$-axis component of the damper winding flux linkages
Ψ_{d1}	$d1$-axis component of the armature flux linkage
Ψ_{q1}	$q1$-axis component of the armature flux linkage
T'_{do}	$d1$-axis transient open circuit time constant
T'_{qo}	$q1$-axis transient open circuit time constant
T''_{do}	$d1$-axis sub-transient open circuit time constant
T''_{qo}	$q1$-axis sub-transient open circuit time constant
R_s	Stator resistance
X_{ls}	Leakage reactance
X''_{d1}	$d1$-axis sub-transient reactance
X''_{q1}	$q1$-axis sub-transient reactance
X'_{d1}	$d1$-axis transient reactance
X'_{q1}	$q1$-axis transient reactance
X_{d1}	$d1$-axis synchronous reactance
X_{q1}	$q1$-axis synchronous reactance
E_{fd}	Field voltage
H	Inertia constant
T_e	Electrical torque
T_m	Mechanical torque
R_D	Droop constant of the governor
P_{ref}	Load reference set-point or turbine reference set-point
P_{sv}	Steam valve output
T_{sv}	Steam valve time constant
T_{CH}	Time constant of the steam chest
V_t	Terminal voltage of the SG-DG
K_E	Exciter gain
T_E	Exciter time constant
K_A	Amplifier gain
T_A	Amplifier time constant
V_{in}	Voltage regulator input
V_R	Voltage regulator output
r_f	Per phase resistance of the filter
l_f	Per phase inductance of the filter
c_f	Per phase capacitance of the filter
r_t	Per phase resistance of the interfacing transformer
l_t	Per phase inductance of the interfacing transformer
C_{dc}	DC link capacitor
V_{dc}	Voltage across C_{dc}
I_{cap}	Current through C_{dc}
I_{pv}	Output current from the photovoltaic array
I_{dc}	DC link current

$\boldsymbol{e}_{\boldsymbol{3abc}}$	*abc* frame components (3-Φ instantaneous values) of the VSI output voltage
$\boldsymbol{v}_{\boldsymbol{3fabc}}$	*abc* frame components (3-Φ instantaneous values) of the filter voltage
$\boldsymbol{v}_{\boldsymbol{3abc}}$	*abc* frame components (3-Φ instantaneous values) of the bus B3 voltage
$\boldsymbol{i}_{\boldsymbol{3fabc}}$	*abc* frame components (3-Φ instantaneous values) of the filter current
$\boldsymbol{i}_{\boldsymbol{3abc}}$	*abc* frame components (3-Φ instantaneous values) of the PV-DG current
$\boldsymbol{E}_{\boldsymbol{3abc}}$	p.u *abc* frame components of the VSI output voltage for $d3$-$q3$ conversion
$\boldsymbol{V}_{\boldsymbol{3fabc}}$	p.u *abc* frame components of the filter voltage for $d3$-$q3$ conversion
$\boldsymbol{V}_{\boldsymbol{3abc}}$	p.u *abc* frame components of the bus B3 voltage for $d3$-$q3$ conversion
$\boldsymbol{I}_{\boldsymbol{3fabc}}$	p.u *abc* frame components of the filter current for $d3$-$q3$ conversion
$\boldsymbol{I}_{\boldsymbol{3abc}}$	p.u *abc* frame components of the PV-DG current for $d3$-$q3$ conversion
R_f	p.u resistance of the filter
X_f	p.u inductive reactance of the filter
X_{Cf}	p.u capacitive reactance of the filter
R_t	p.u resistance of the interfacing transformer
X_t	p.u inductive reactance of the interfacing transformer
E_{d3}	$d3$-axis component of the VSI output voltage
V_{fd3}	$d3$-axis component of the voltage across the filter capacitor
V_{d3}	$d3$-axis component of the bus B3 voltage
I_{fd3}	$d3$-axis component of the filter current
I_{d3}	$d3$-axis component of the PV-DG current
E_{q3}	$q3$-axis component of the VSI output voltage
V_{fq3}	$q3$-axis component of the voltage across the filter capacitor
V_{q3}	$q3$-axis component of the bus B3 voltage
I_{fq3}	$q3$-axis component of the filter current
I_{q3}	$q3$-axis component of the PV-DG current
K_p	PLL proportional gain
K_i	PLL integral gain
φ_{PLL}	PLL intermediate state variable
G	Solar irradiance (W/m^2)
T	Surface temperature (^{0}C)
q	Electron charge (Coulombs)
A	Diode quality factor
K	Boltzmann's constant
N_{pa}	Number of parallel rows in each module
N_{se}	Number of series cells in each parallel row of a module
I_{sc}	Short circuit current of the module (A)
I_{scn}	Short circuit current of the module at nominal conditions (A)
K_I	Temperature coefficient of the short circuit current (A/°C)
T_n	Nominal temperature (°C)
V_{oc}	Open circuit voltage of the module (V)
V_{ocn}	Open circuit voltage of the module at nominal conditions (V)
K_V	Temperature coefficient of the open circuit voltage (V/°C)
I_0	Dark saturation current of the diode (A)
R_{se}	Equivalent series resistance of each module (Ω)

R_{sh}	Equivalent shunt resistance of each module (Ω)
I_{ph}	Photo generated current of each solar cell (A)
G_n	Nominal irradiance (W/m^2)
N_{pp}	Number of parallel rows of modules in the PV array
N_{ss}	Number of series modules in each parallel row of the PV array
V_i^D	*D*-axis component of *i*th bus voltage
V_i^Q	*Q*-axis component of *i*th bus voltage
I_{li}^D	*D*-axis component of *i*th bus load current
I_{li}^Q	*Q*-axis component of *i*th bus load current
R_{li}	Resistance of the load at *i*th bus
X_{li}	Reactance of the load at *i*th bus
I_{mi}^D	*D*-axis component of *i*th bus induction motor load current
I_{mi}^Q	*Q*-axis component of *i*th bus induction motor load current
E_{mi}^D	*D*-axis component of the induced voltage of the induction motor at *i*th bus
E_{mi}^Q	*Q*-axis component of the induced voltage of the induction motor at *i*th bus
R_{mi}	Stator resistance of the induction motor at *i*th bus
X'_{mi}	Transient reactance of the induction motor at *i*th bus
X_{mi}	Synchronous reactance of the induction motor at *i*th bus
T_{mi}	Transient open circuit time constant of the induction motor at *i*th bus
S_{mi}	Slip of the induction motor at *i*th bus
H_{mi}	Inertia constant of the induction motor at *i*th bus
T_{emi}	Electrical torque of the induction motor at *i*th bus
T_{mmi}	Mechanical torque of the induction motor at *i*th bus
I_{ij}^D	*D*-axis component of the line current between buses *i* and *j*
I_{ij}^Q	*Q*-axis component of the line current between buses *i* and *j*
R_{ij}	Resistance of the line between buses *i* and *j*
X_{ij}	Inductance of the line between buses *i* and *j*
NB	Number of buses in the micro-grid
I_i^D	*D*-axis component of the generator injected current at *i*th bus
I_i^Q	*Q*-axis component of the generator injected current at *i*th bus
X_{Ci}	Shunt capacitance at *i*th bus
$\boldsymbol{X}$	Continuous-time state vector
N	Number of states
$\boldsymbol{V}$	Input vector
n_{ip}	Number of inputs
$\boldsymbol{Y}$	Output vector
n_{op}	Number of outputs
P_{G3}	Active power injected at bus B3 by the PV-DG
V_1	Voltage at generator bus B1
V_3	Voltage at generator bus B3
t_s	Present time moment (seconds)
T_s	Sampling time of the controller (seconds)
t	Continuous-time (seconds)
k	Discrete-time (samples)

k_i	Present sample
N_p	Length of the prediction horizon in samples
N_c	Length of the control horizon in samples
$\boldsymbol{U}$	Vector with incremental control trajectories within the control horizon
$\boldsymbol{Y_E}$	Vector with micro-grid outputs predicted within the prediction horizon
$\boldsymbol{U_{opt}}$	Vector with optimal incremental control trajectories within the control horizon
$\boldsymbol{W}$	Vector with future set-points for the outputs within the prediction horizon
$\boldsymbol{R}$	Positive definite weight matrix on input increments
P_{G1}	Active power output of the SG-DG
Q_{G1}	Reactive power output of the SG-DG
Q_{G3}	Reactive power output of the PV-DG
$P_{G1,min}$	Minimum active power of the SG-DG
$P_{G1,max}$	Maximum active power of the SG-DG
$Q_{G1,min}$	Minimum reactive power of the SG-DG
$Q_{G1,max}$	Maximum reactive power of the SG-DG
$P_{G3,min}$	Minimum active power of the PV-DG
$P_{G3,max}$	Maximum active power of the PV-DG
$Q_{G3,min}$	Minimum reactive power of the PV-DG
$Q_{G3,max}$	Maximum reactive power of the PV-DG
V_{mppt}	Voltage of the PV module corresponding to the maximum power point
$P_{3,nom}$	Nominal output of the PV-DG
R_{D3}	Droop constant of the PV-DG
ε	Error between the predicted and actual response of the micro-grid output
y_{actual}	Actual response of the micro-grid output
$y_{predicted}$	Predicted response of the micro-grid output
$\boldsymbol{X_{ref}}$	Reference trajectory of the state vector
$\boldsymbol{V_{ref}}$	Reference trajectory of the input vector
$\boldsymbol{Y_{ref}}$	Reference trajectory of the output vector
$\Delta \boldsymbol{Y_p}$	Vector with forced response of the micro-grid within the prediction horizon
$\boldsymbol{Y_p}$	Vector with natural response of the micro-grid within the prediction horizon
$\boldsymbol{V_{opt}}$	Optimal value of the input vector
$\boldsymbol{V_{tail}}$	Tail of the input vector
c_i	ith coefficient in the special function network of N functions
o_i	ith special function in the special function network of N functions
l_i	ith Laguerre function in a Laguerre network of N functions
p	Real pole of the Laguerre network
$\boldsymbol{Lag}$	Laguerre network of N functions
N_s	Number of special functions to approximate the control trajectory of sth input
$\boldsymbol{Lag_s}$	Laguerre network to approximate sth input
η_S	Coefficient vector of sth input
$\boldsymbol{L}$	Complete Laguerre matrix

$\boldsymbol{\eta}$	Complete coefficient vector
p_1	Complex pole of the Kautz network with N functions
$\overline{p_1}$	Complex conjugate pole of the Kautz network with N functions
K_i	ith Kautz function in a Kautz network of N functions
$\boldsymbol{Kat}$	Kautz network of N functions
$\boldsymbol{Kat}_S$	Kautz network to approximate sth input
$\boldsymbol{K}$	Complete Kautz matrix
$\boldsymbol{O}$	Complete special function matrix
$\boldsymbol{w}$	Vector having set-points for each output in the output vector
$\boldsymbol{r}$	Vector having penalties on each input increment in the incremental input vector
$\boldsymbol{R_w}$	Penalty matrix on input increments
$\boldsymbol{\eta_{opt}}$	Optimal coefficient vector
C	Complexity factor
N_{sg}	Number of synchronous generators
N_{pv}	Number of PV generators
$\boldsymbol{\Upsilon}$	Disturbance vector
$\widehat{\boldsymbol{\Upsilon}}$	Estimated disturbance vector from the compensator
$\boldsymbol{\alpha}$	Gain matrix of the compensator
B9	Bus number 9
B10	Bus number 10

Chapter 1
Micro-grid Introduction and Overview

Abstract The chapter provides a detailed explanation about the reasons for the evolution of micro-grids. The conventional power system components, its architecture, and the challenges it poses in the modern-day power sector are discussed in Sect. 1.1. The concept of distributed generator (DG) and the typical components involved in a DG are explained in the Sect. 1.2. The role of DG in overcoming some of the challenges posed by the conventional power systems is explained in Sect. 1.3. However, a single DG always has its demerits when operated in either standalone or grid-connected modes. These demerits are discussed in Sect. 1.4. The concept of micro-grid, its definitions, and its ability to overcome the demerits of a single DG are discussed in Sect. 1.5. Section 1.6 provides a detailed explanation about different components involved in the modern-day micro-grid. The advantages, challenges, and operational modes of a micro-grid are explained in Sects. 1.7, 1.8, and 1.9, respectively.

Keywords Distributed Generation · Distributed Storage · Grid-connected mode · Micro-grid · Standalone mode

1.1 Conventional Power Systems Review

The conventional power system is a setup where the power generation happens in bulk at concentrated locations called power stations. The power is then transmitted over long distances at high voltage levels through transmission lines. The transmitted power is then distributed to customers at the distribution level at low voltage levels. This kind of setup shown in Fig. 1.1 served the power consumers for over a century. The conventional power system relied heavily on non-renewable sources, mainly coal, oil, and natural gas, as raw materials for the generation. Boiler, multi-stage steam turbines with different pressure levels, and a synchronous generator are the basic machinery or traditional technology used for generation. The generation will be owned by the generation companies called GENCO's, the transmission will be owned by transmission companies called TRANSCO's, and distribution will be owned by distribution companies called DISCOM's. These companies can be either

P. Vidyasagar and K. Shanti Swarup, *Design and Development of Model Predictive Primary Control of Micro Grids*, Springer Tracts in Electrical and Electronics Engineering, https://doi.org/10.1007/978-981-19-5852-6_1

fully government-owned or private-owned or a combination of government-owned and private-owned. One important aspect to note here is the negligible presence of customers participating in the generation activities. They are solely there to consume the power produced by the GENCO's in the enormous power stations. This conventional power system setup is the main driving force for the technological advancements in the transport sector, small-, medium-, and large-scale industrial sectors, defence sector, and agricultural sector over the past 100 years.

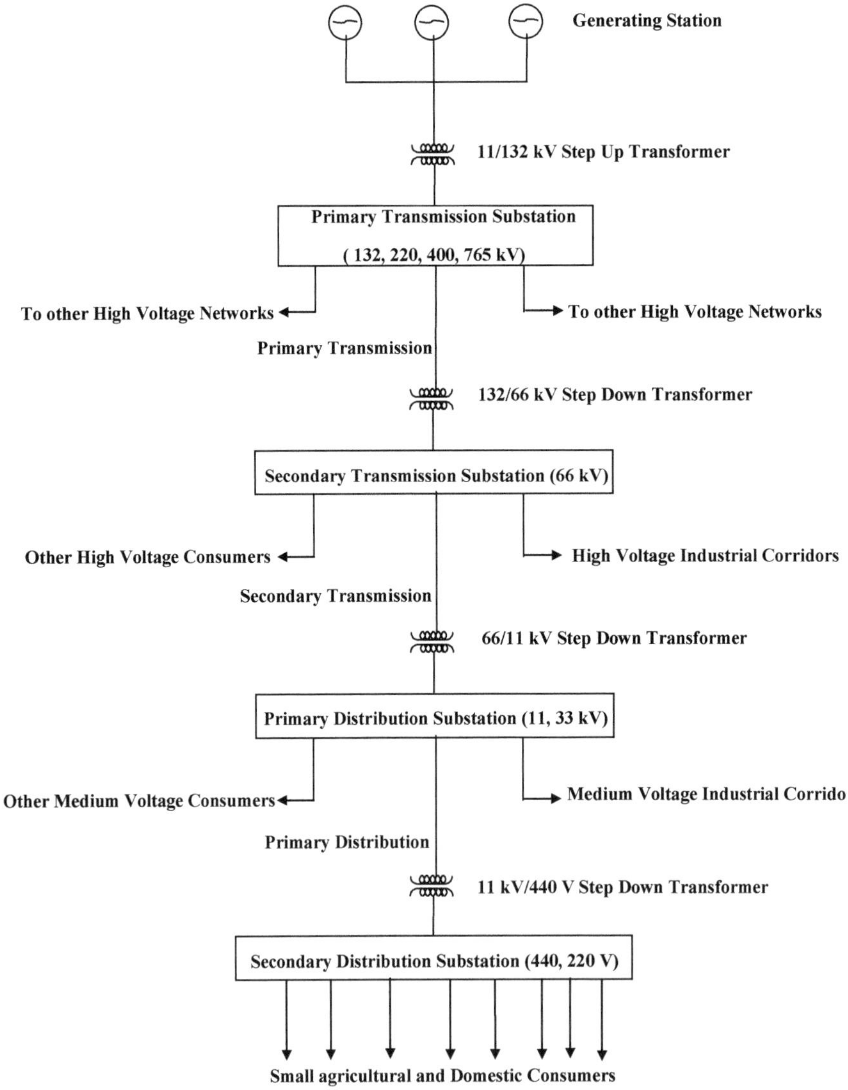

Fig. 1.1 Typical conventional power system structure

But it is well known that every industry or aspect of a society will experience a change in trend. In some sectors, these changes happen rapidly, and in some, it happens after decades. In the case of power systems, it's happening after 100 years of conventional setup. The change in trend is happening at all levels, i.e. generation, transmission, and distribution. The primary trend we are interested in is on the generation side, called the distributed generation (Lopes et al. 2007; Dondi et al. 2002). Before going into the distributed generation, it is first essential to understand the demerits of the conventional power systems. The following are the major ones with many other minor ones:

- **Shortage of non-renewable sources**: Coal, oil, and natural gases are non-renewable. Their heavy consumption over the past century naturally leads to their shortages in many countries.
- **Unfair sources distribution**: The sources are not uniformly distributed across the countries. This is causing unfair advantage to some countries and a significant disadvantage to some other countries especially developing countries which depend on the other countries having these sources.
- **Carbon footprint and climatic effects**: Perhaps the most devastating effect of the conventional generation systems is the carbon emissions and their subsequent effect on climate change. A grave concern that is making headlines these days in developed countries, climate change, is a reality that should be addressed.
- **Low energy efficiencies**: The energy efficiency of a conventional power plant fed with coal, oil, or natural gas is very low. It is below 50% in most cases. Generating a bulk amount of power at such low efficiencies is a disadvantage in energy conservation.
- **Losses due to transmission over long distances**: The power transmission over long distances results in losses, thereby denting the available power.
- **Gross negligence of local resources**: Wind, solar, tidal, biomass, and geothermal sources are abundant in many places on the earth. They may not be concentrated in large amounts in a single location but are available in a quantity that can meet the power demands of the respective local areas. This requires decentralization of generation or, in other terms, a provision for integrating these sources into the grid. At this stage, by a grid, we mean the conventional large-scale power system setup explained above.
- **Cost involvement in the expansion of conventional power systems**: As the power demand continuously increases, so is the necessity to build new power plants and transmission lines. This is a very hectic task in terms of the location's geography, the economic ability of the local governments, and the environmental impacts.

Apart from the above significant concerns, conventional power generation faces many minor issues local to the areas of their presence. Given the above problems, now we can understand why the new trend of distributed generation is gaining importance and rapidly increasing worldwide.

1.2 Concept of Distributed Generation

The traditional generation is generally concentrated at a few locations. The location of the power plants, availability of raw energy material, and employed human resources are all concentrated in very few places. But in nature, many energy resources are distributed. These distributed energy resources (DERs) are the sources of energy spread across the world and are close to load centres or consumer areas. The basic idea of a distributed generator (DG) is to utilize these DERs to generate a small amount of power to meet the local needs and inject surplus power into the utility system (Lopes et al. 2007; Dondi et al. 2002). The main feature of the DGs is their smallness. Their ratings are in kW and usually less than 500 kW. They are connected at the traditional power system's medium voltage or low voltage levels, which is the distribution level as shown in Fig. 1.2. Any person, community, or industry who can extract energy from the local DERs can own the respective DG. But he should adhere to the grid codes when integrating and operating the DG with the grid. The concept of DG allows the consumers to become prosumers where they can produce the energy and sell it to the grid during the surplus times and consume from the grid during the deficit times.

The typical constituents of a DG are shown in Fig. 1.3. It consists of a primary input energy source out of which energy is extracted to convert it into electrical energy. The interfacing medium will convert the primary energy into electrical energy. The supervisory system will be responsible for the data collection through measurements, state estimation, and close monitoring of the DG operation. The protection and control systems are responsible for the safe, secure, and reliable operation of the DG with adherence to grid standards. The energy resource for a DG can either be fossil fuel or renewable. Diesel, natural gas, biogas, solar energy, wind energy, tidal energy, and geothermal energy are the dominant distributed energy resources around the globe. The standard DG technologies that utilize these resources include:

- Photovoltaic (PV) modules
- Wind turbines coupled with induction generators
- Reciprocating engine generators
- Micro-turbines and gas turbines coupled with synchronous generators
- Fuel cell technologies with gasoline
- Combined heat and power (CHP) equipment at a small scale to recover the waste heat from the conversion of primary fuel to electric power to locally utilize it for heating purposes, thereby increasing the energy efficiency
- Biogas digesters.

Apart from the above technologies, the distributed generation also involves different storage technologies for electricity, heat, and gas, including batteries, flywheels, ultra-capacitors, etc. Combining storage with distributed generation provides peak power and ride-through capabilities during disturbances in the grid. In literature, even a load that can be controllable can also be considered as a DG under some particular scenarios like demand-side management.

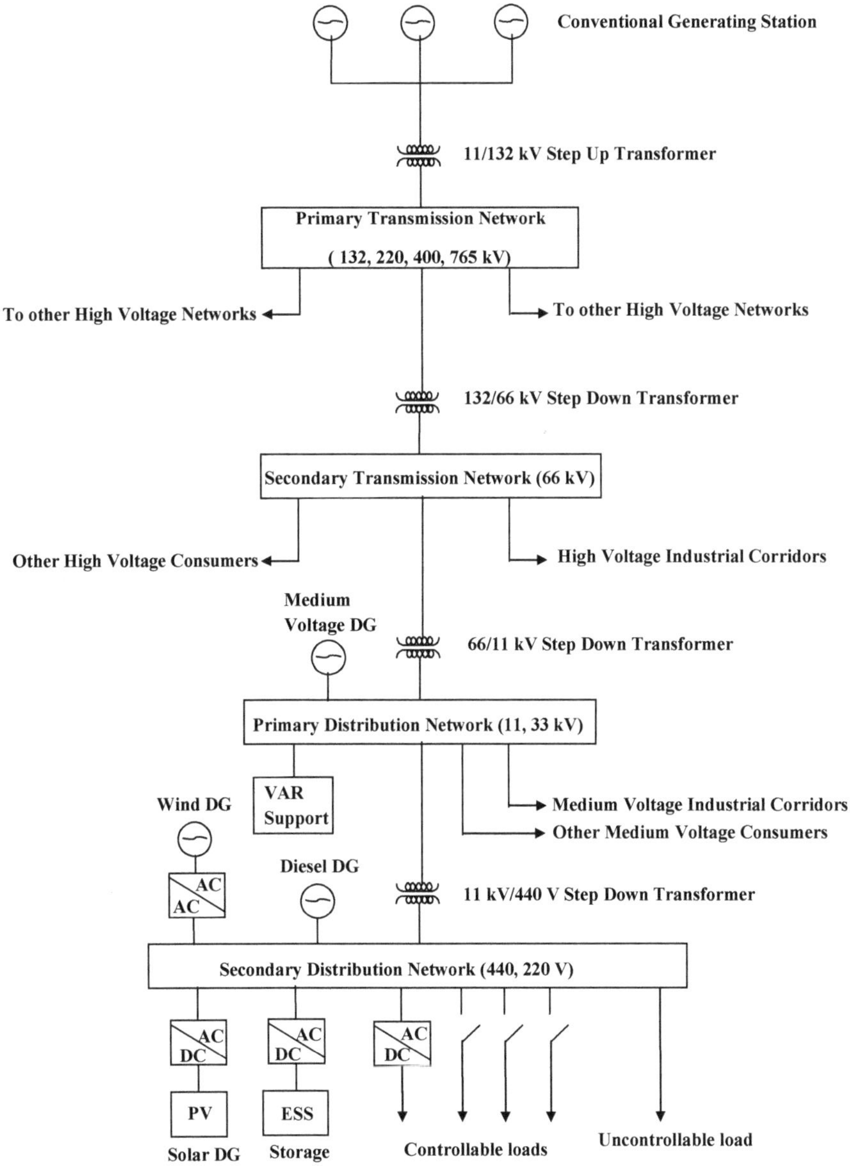

Fig. 1.2 Power system with integrated distributed generation

1.3 Necessity of Distributed Generation

The conventional generation in a power system is no longer a viable option for the future for various reasons explained above. The distributed generation can address

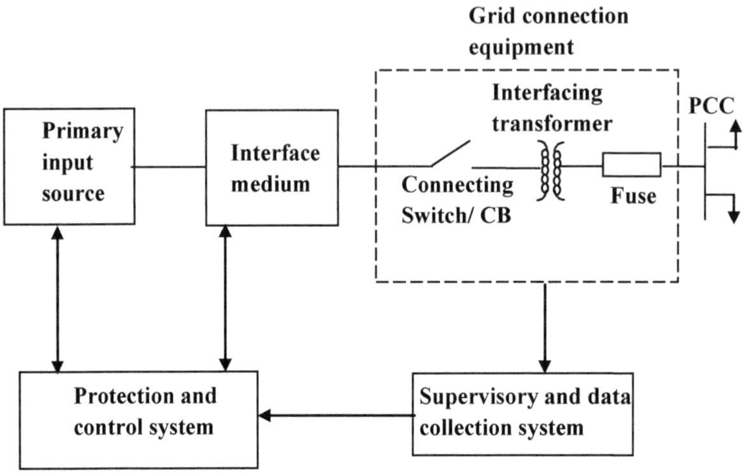

Fig. 1.3 Typical constituents of a DG

many concerns plaguing the conventional generation and provide some extra benefits that are not in the curriculum of the traditional power plants. The distributed generation has become a reality and an integral part of the grid and consumer trust. The potential benefits of the distributed generation are as follows (Lopes et al. 2007; Dondi et al. 2002):

- **Improved energy efficiency using small-scale CHP's**: A small-scale combined heat and power plant can significantly improve the energy efficiency of fossil fuel generation. This is because the waste heat from the electricity generation can be captured, stored, and used for local heating purposes. In a conventional power system, this is very hard to achieve as the distance between power plants and the consumers is very long.
- **Less carbon footprint**: It is the commitment of the developed countries to reduce greenhouse gas emissions, and each country has its target to meet in terms of reduced carbon footprint. For this, these countries must focus on renewable and environmentally friendly energy sources. Ideally, distributed generation is more suitable for renewable sources than the conventional concentrated plants for electricity production. This is due to the basic fact that, unlike coal or gas, renewables cannot be transported to focused locations.
- **Power Quality Improvement**: Generation close to load can improve the reliability and quality of the power. One of the extra benefits of the distributed generation is that they can be employed solely for power quality improvement.
- **Price reduction of the electricity**: With the penetration of more and more renewable DGs into the grid, the price of the electricity becomes cheaper as the raw material cost, transmission cost, and other variable operating and maintenance costs of the traditional power plants are reduced.

- **Participation of consumers in the electricity markets**: This is one area that provides potential business opportunities in power production for individuals, start-ups, small-scale and medium-scale power producers. Earlier, electricity generation lies in the hands of governments and prominent industrialists as the capital costs are very high for the traditional power plants.
- **Energy-saving through loss reduction**: As the generation is close to loads, the transmission line losses will be reduced, thereby causing energy savings.
- **Reducing the expansion burden on governments**: One of the biggest challenges to the developing countries is that their power demand increases with time, whereas the electricity infrastructure is not. This is due to the massive amount of capital involved in expanding generation and transmission. Since most developing countries rely heavily on governments to develop electricity infrastructure, it is becoming a significant burden. Distributed generation is one efficient solution that can address this problem.
- **Ensuring a reliable future**: Climate change, shortage of non-renewable sources, and increasing power demand are the real problems the future generations will face if the conventional power generation continues as it is. To keep future generations away from these problems and provide a reliable future, distributed generation is one of the many requirements that the power systems need to adopt.

1.4 Single DG Challenges

With every new technology, there come new challenges. DGs provide effective solutions for utilizing the local resources, reducing the traditional system expansion, losses reduction, and emission reduction. But the penetration of a large number of DGs causes several control and operational challenges for the safe operation of the grid (Lopes et al. 2007; Dondi et al. 2002).

- **Power conversions**: Some DG sources are DC in nature. Some other sources are AC, but the frequency is very high. When integrating a DG to the grid, it must follow grid code which generally includes the prescribed frequency range, voltage range, harmonic content, and other power quality specifications. To achieve all the above requirements, the DGs require power electronic converters with a local control system designed to achieve these specifications from the grid. The DG control system should be capable of the following:
- Make sure the injected power into the grid is at grid frequency and the voltage at the interconnection, also called point of common coupling (PCC), is not disturbed.
- It should make sure that the DG injects the committed active and reactive powers into the grid irrespective of changes happening on the grid. This mode is also called P-Q mode of operation.
- The control system should also make sure that, in case of disconnection from the grid, the DG bus should act as a PV bus supplying the local load at rated voltage and allowable frequency of the load.

- The control system should make sure that the disconnection and connection of the DG with the grid must be smooth.
- **Less inertia**: The traditional power systems are generally equipped with large synchronous generators for the generation. Onc of the significant advantages is that the grid exhibits large inertia, which can store energy and, if required, can be extracted out of it at the expense of speed or, in turn, frequency during the critical disturbances. But if the grid is penetrated with a large number of inverter-based DGs, then its inertia will be less, and during the disturbances, it exhibits large fluctuations in the frequency.
- **Intermittency of the sources**: One of the disadvantages of renewable energy-based DGs is that the sources are intermittent. Solar power and wind power are not available throughout the day. Even during the time they are available, they experience large fluctuations. This, in turn, makes the DGs non-dispatchable and highly volatile. In the case of large penetration of DGs, this volatile nature of DG outputs leads to instability of the grid.
- **Less reliable for day-ahead market analysis**: The DGs can inject the surplus power into the grid and make money out of it. But, generally, the day-ahead market requires generation bids from the generators to make them participate in the unit commitment studies. This is a challenging aspect for the distributed generators as they cannot plan the generation. Hence, they are limited to real-time markets and have a minimal choice in terms of pay for power injection.
- **Difficult to supervise**: It is challenging for the system operator to manage many DGs integrated into the grid. Also, the DG maintenance and operation requires experts, which will burden the owners of the small-scale DGs.
- **Frequent plug-and-play impact**: Sometimes, these DGs are frequently disconnected and reconnected due to various technical, economic, and political reasons. It will result in fluctuations in frequency and voltages in the distribution system if a large number of connections and disconnections are happening in a brief span of time, affecting the grid's stability.
- **Congestion**: Large numbers of DGs are causing congestion issues in the distribution system.
- **Frequency and voltage regulation**: A single DG connected to the grid cannot participate in frequency and voltage regulation.

1.5 Concept of Micro-grid and Definitions

From the previous section, we understood that the integration of many individual DGs is challenging, complex to control with strict adherence to grid codes, and has severe impacts on the network in terms of power quality and stability if not correctly operated. At the same time, due to their smallness and source intermittency, the local demand cannot be met reliably and still require a significant contribution from the grid. This defeats the purpose of the distributed generation most of the time when they are disconnected from the grid. Another important aspect is that the individual DGs cannot participate in markets and cannot provide ancillary services. Hence, there is a new paradigm for integrating DGs with the grid and their effective operation when disconnected from the grid. The concept of micro-grid is the most popular among the available paradigms, so far the most successful one and the most appealing one.

The micro-grids were first introduced to the power systems in the 2001 IEEE PES WM panel led by Bob Lasseter, followed by his conference paper and the CERTS report in 2002 (Lasseter 2001, 2002; Ramakumar 2001). The basic idea of micro-grid is to aggregate different DGs, loads both controllable and uncontrollable, and storage elements into an independent network that can be operated autonomously and connected to the grid's distribution network as shown in Fig. 1.4. It provides a reliable connection of the local DERs to meet the local demands efficiently and inject surplus power into the grid (Olivares et al. 2014). The individual DGs, controllable loads, and storage elements are very small to deal with the source intermittency, causing fluctuations in the power production and rapid changes in the local loads. But as a single network, they can provide certainty in production and consumption. They can participate in market operations and provide ancillary services as a network. The concept of micro-grid allows the DGs to participate in the frequency regulation and voltage control. From the utility point of view, a micro-grid is a single controllable entity that can respond to the central control commands and participate in the electricity markets, frequency control, and voltage control. In other words, it acts as a single good citizen that obeys the grid codes. From the local customer point of view, it can be viewed as a reliable generation and distribution network that can be trusted for uninterrupted power supply irrespective of grid condition, at low prices, and high power quality. It provides a great platform for the economy for many prosumers during the grid's peak load and emergency conditions.

There is no strict mandate on what types of loads and their geographical area a micro-grid can serve. The different communities a micro-grid can serve vary from large urban residential areas to small isolated rural residential areas that are cut off from the grid. The micro-grids serve gated communities, academic institutions, and public institutions. Of all the communities, the industrial parks are the most important ones. The micro-grid serving an industrial corridor dramatically reduces the losses, provides less carbon footprint businesses, decreases the peak demand on the grid, and decreases the energy dependency. At this stage, let's see some of the definitions for micro-grid. A strict technical definition for micro-grid is still under discussion among the electrical forums. A general definition for the micro-grid can be given as a cluster of DGs, energy storage systems, and loads operated in coordination to reliably supply electricity, connected to the power system at the distribution level at a single point of connection (PCC). Let us look at the micro-grid definitions from different technical forums:

As per IEEE standard 2030.7, a micro-grid is a group of interconnected loads and distributed energy resources with clearly defined electrical boundaries that act as a single controllable entity with respect to the grid and can connect and disconnect from the grid to enable it to operate in both grid-connected or island modes.

As per CIGRE, micro-grids are sections of electricity distribution systems containing loads and DERs (such as DGs, storage devices, or controllable loads)

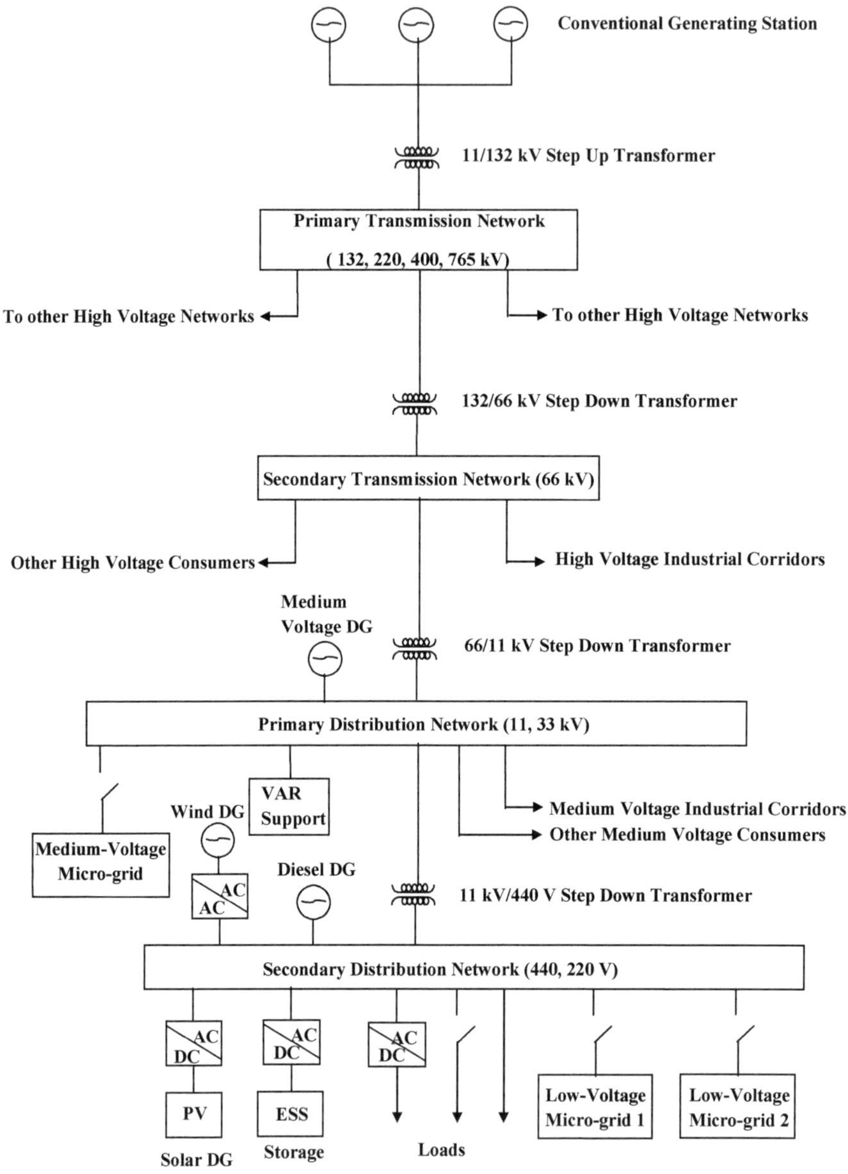

Fig. 1.4 Power system with integrated distributed generation and micro-grids

that can be operated in a controlled, coordinated way, either while connected to the main power network and/or while islanded.

In both these definitions, one thing we can notice is that, the micro-grid differs from the distribution network in two aspects: (i) It can operate as an islanded autonomous grid increasing the reliability and resilience. (ii) For an upstream network, it is single entity that can be controlled.

1.6 General Constituents of a Micro-grid

It is not that the concept of micro-grid is something very new or out of box thought. This kind of small autonomous grids existed in rural and remote communities that were cut off from the main grid due to infeasible technical, environmental, and economic reasons. But one thing the present-day micro-grid differs from these earlier autonomous grids is the constituents present in it. Non-renewable fossil fuels are the most common generation sources in the earlier autonomous grids due to their flexibility and scalability. But the, present-day micro-grids constitute much more than the fossil fuel-based generation as shown in Fig. 1.5. The priority of the micro-grids is to incorporate greener technologies for the generation like solar, wind, hydro, and hydrogen. Apart from the generation, the micro-grids differ from the earlier autonomous grids in controllable loads and energy storage. Hence it is fascinating to know the fundamental constituents of a modern micro-grid (Kroposki et al. 2008; Katiraei et al. 2008).

- **Distributed generation**:

(a) **Based on Interface**:
Let us first look at the types of DGs in a micro-grid. There are two types of DGs in a micro-grid based on the interface. One is a group of DGs connected to the micro-grid through rotating machines. We call them conventional DG units as the equipment in these DGs resembles the conventional generators in a power plant. The other is a group of DGs that are interfaced and connected to micro-grid via power electronic converters. These are called electronically interfaced DGs. A conventional DG unit consists of either a synchronous generator driven by a reciprocating engine or an induction generator driven by a fixed speed wind turbine as shown in Fig. 1.6. The primary energy source can be diesel, natural gas, biogas, wind, hydro energy, tidal, or wave energy, to name a few. The rotating machine converts the primary fuel energy into electrical energy and acts as an interface to the micro-grid injecting power at specified frequency and voltage. In an electronically interfaced DG unit, there can be a single DC-AC inverter or an AC-DC converter followed by a DC-AC inverter or a DC-DC converter followed by a DC-AC inverter as shown in Fig. 1.7. The power electronic converters act as coupling converters that converts the electrical energy from the primary energy sources into AC electrical energy with

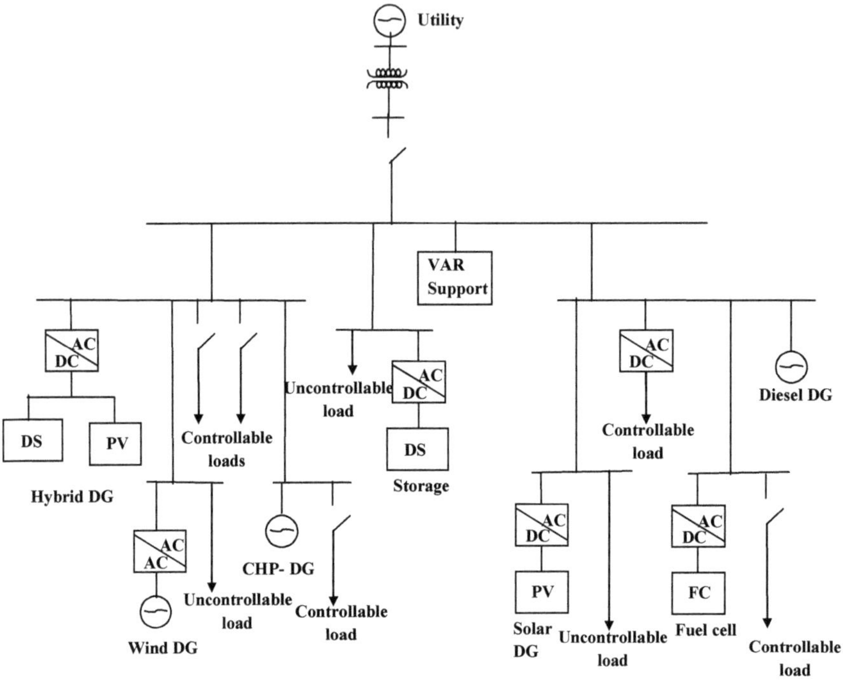

Fig. 1.5 Typical constituents of a modern-day micro-grid

desired specifications. They act as an interface for the DG unit to connect with the micro-grid and form a basic control layer that can control the frequency and voltage of the power injected into the grid. The converters can be unidirectional or bidirectional based on storage availability, which will be discussed in a while. The input power can be DC or AC at a fixed or variable frequency for the source side converter. For the micro-grid side of the converter, the output power is AC with a frequency of 50 Hz or 60 Hz. The input side technologies for these electronically interfaced DGs are PV modules, micro-turbines fuelled by non-renewable sources, induction generators with variable speed wind turbines, fuel cells, to name a few.

(b) **Based on dispatch**:

The type of DG units can also be classified in terms of their ability to dispatch power. This, in turn, means their ability to control their output power as per the external command signals from the supervisory control systems. There are dispatchable DG units and non-dispatchable DG units. For example, a reciprocating engine connected to a synchronous generator is a dispatchable DG unit. The primary energy source can be diesel, natural gas, biogas, or any other fossil fuel. The DG unit is equipped with a governor for speed control and fuel input control. The DG unit is also equipped with voltage regulators that can control the internal voltage of the synchronous generator. These two control systems

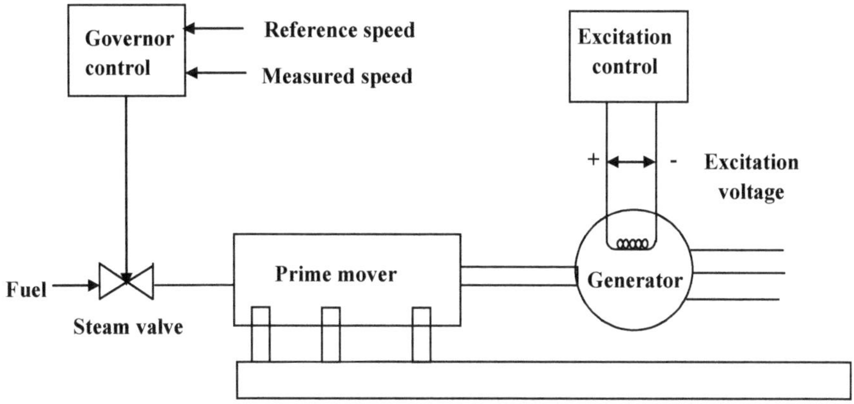

Fig. 1.6 Typical constituents of a DG with rotating interface

control the active and reactive power outputs of the DG unit as per dispatch strategy. On the other hand, the output power of a non-dispatchable DG unit is generally not controllable as per our needs. For example, a PV-DG unit output depends on solar irradiation, which is not in our hands. Hence in such types of DG units, the DG is controlled in such a way that the maximum power possible is extracted based on the maximum power point tracking (MPPT) algorithm.

- **Distributed Storage**:
  Storage is one of the most important aspects of the micro-grid, especially in the islanded or autonomous mode. Micro-grid dominated by renewable sources is highly volatile in terms of power generation. Many times, the power generation cannot match the local load. In such cases, the storage will improve the reliability of the micro-grid by trying to match the power generation with the load demand. It stores the surplus power in the micro-grid during high amounts of source availability. It acts as a bridge that can meet the micro-grid power requirements during a shortage in source availability. Its capacity is determined either by using its energy density or power density. The advantages of storage in a micro-grid are:

  (a) The power generation from a DG need not respond to load fluctuations. It can still produce the constant and stable output, which many times corresponds to maximum power point for renewable sources and unit commitment output for non-renewable sources.
  (b) It provides ride-through capability during source intermittencies of solar, wind, and hydropower sources.
  (c) The third important contribution of storage is that it makes the non-dispatchable unit a dispatchable one because the storage provides the leverage to increase or decrease the generation from the DG unit when performing unit commitment studies.

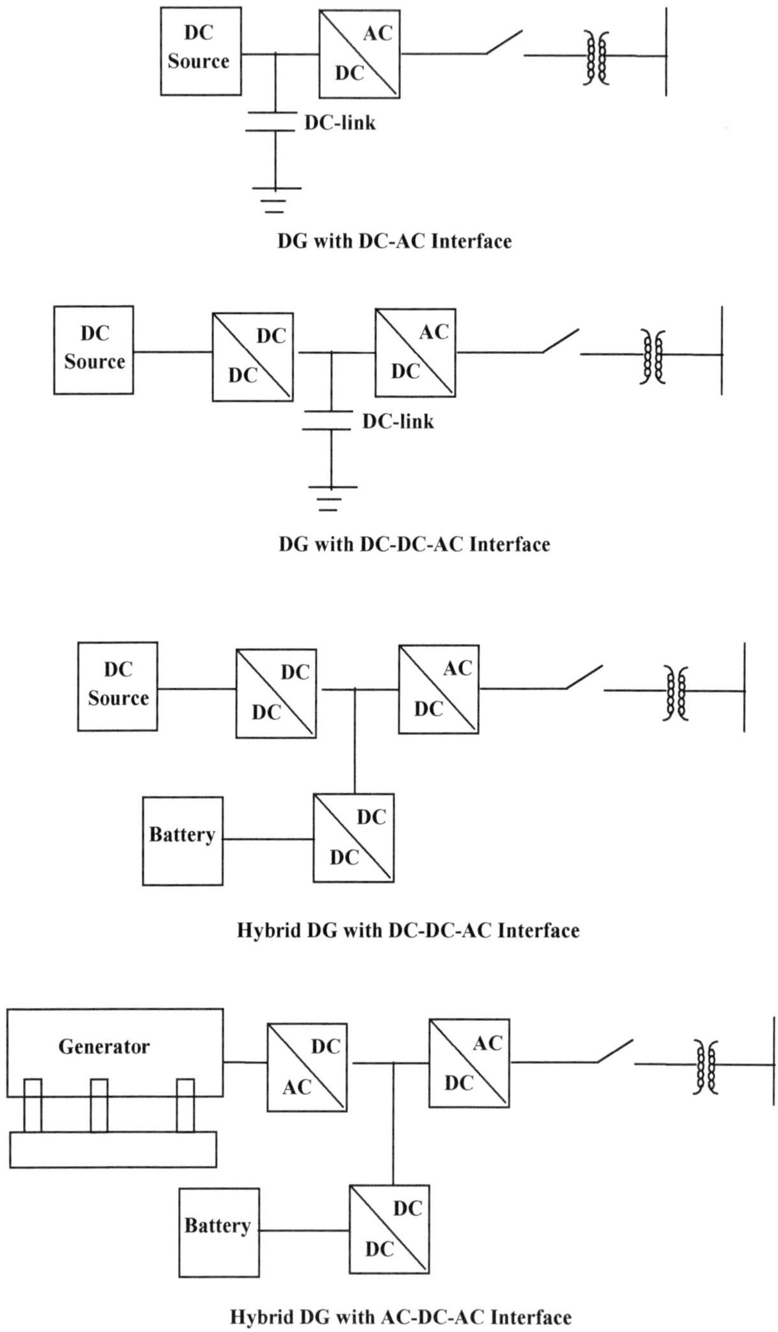

Fig. 1.7 Typical constituents of a DG with power electronic interface

(d) The presence of energy storage is also beneficial to the distribution network to which the micro-grid is connected. It provides reserves, outage ride-through capabilities, damping peak demand, and counters the sudden load disturbances in the distribution systems.

The most common types of energy storage systems employed in a micro-grid are:

(a) **Batteries**: Batteries store electrical energy in the form of chemical energy. The output power from a battery is DC. Hence we need power electronic devices to convert the energy from and to the AC. If the battery system is a separate entity, bidirectional converters must convert AC to DC while charging and DC to AC while discharging. If it is a part of the DG unit whose source is DC, for example, PV modules, it can be directly connected across the DG output terminals as part of the DC-link capacitor. In this case, a single inverter for DC to AC conversion is enough, or a DC-DC-AC system is used. If it is a part of DG units whose output is AC, for example, wind, then it is connected at the DC terminals of an AC-DC-AC converter system via a DC-DC converter.
(b) **Super capacitors (Ultra-capacitors)**: They have extremely high power density, cycling capability, and very fast response to the power fluctuations in the system.
(c) **Flywheel systems**: A new trend in micro-grid storage, the flywheel technology is very helpful to deal with the power interruptions in the grid due to fast response time. Generally, the flywheel is connected to a motor that serves the mechanical energy to the flywheel. When generation is required, the flywheel acts as a mechanical source to a generator connected to the same shaft that outputs power through a converter. It is also possible to design a bidirectional system with one machine that is capable of motoring and regenerating operations.

- **Hybrid DGs**: A hybrid DG is a term associated with those generators which have a DG unit with a primary energy source together with a distributed storage unit. The combination of DG and DS makes the whole generator dispatchable.
- **Loads**: A micro-grid consists of both electrical and thermal loads. There are two categories of load classification in micro-grid. The first classification is based on the electric elements present in the load. Impedance loads, constant power loads, constant current loads, loads with power electronic equipment, and thermal loads come under this load classification category. The other classification is controllable and uncontrollable loads. The uncontrollable loads are very sensitive to the power quality parameters, cannot be adjusted to different time frames, and cannot be operated in partial capacities. Their demand has to be met by the micro-grid, whether it is connected to the grid or operating in autonomous mode. The best example would be loads from the emergency services like hospitals. On the other hand, a non-sensitive load can come under controllable loads. These loads can be controlled to reduce the peak demand, deal with power shortages, and deal with disruptions in the grid or micro-grid. They can be scheduled to different time frames to smooth out the load profile of the micro-grid. These loads are the first ones to be shed

when it is required. The controllable loads participate in the demand response strategies conducted by the supervisory control system of the micro-grid.

1.7 Advantages of a Micro-grid

Let us look at the most important advantages a micro-grid can provide to utility and the micro-grid participants, out of which some are prosumers, and some are customers. For a utility:

- From a utility point of view, a micro-grid can be represented as a single prosumer in the unit commitment, optimal power flow, load flow, and expansion studies.
- The micro-grid can address the increasing congestion problems in the utility network.
- The losses in the utility network can be significantly reduced.
- Micro-grids can participate in the voltage and frequency control of the host network.
- The increasing peak demands in the distribution network can be addressed quickly with the micro-grids rather than the conventional power system expansion.
- The huge amount of capital involved in the power system planning and expansion can be saved to a great extent with carefully designed and operated micro-grids.
- Rural and remote areas which cannot be connected to the utility for various reasons can be electrified using micro-grids.
- Reduces the environmental impact.
- Improves the resilience of the power systems to natural and manmade disasters.

From a customer point of view, the micro-grid:

- Improves the power quality and reliability
- Enables them to participate in the markets
- Decreases the price of the power.

1.8 Micro-grid Challenges

As said earlier, every new technology or concept comes with new advantages as well as challenges. The micro-grid is no exception to this statement. The challenges one should address in a micro-grid are:

- The skillset required to operate micro-grid is very specific and hence require professionals with good technical background and operational control capabilities.
- Load and source forecasting is a crucial aspect while performing the schedule and dispatch studies. The uncertainty in generation and demand in a micro-grid is to be considered in these studies.
- Determining the appropriate reserve levels for frequency and voltage control during disturbances in the local load or in the grid is another challenging aspect.

- Designing demand-side management to allow customers to participate in the incentive-based control strategies involving meeting the grid requirements.
- Designing new market models to allow the customers and prosumers to participate in the competitive electrical markets.
- Due to bidirectional power flows in micro-grid, the conventional protection systems need to be reengineered.
- Developing control techniques for power flow control in the grid-connected mode and voltage and frequency control during the islanded mode.
- Developing seamless plug-and-play mechanisms.

1.9 Micro-grid Operational Modes

A micro-grid operates in two modes. One is the grid-connected mode, and the other one is islanded mode. The islanded mode is also called standalone mode or autonomous mode. These two modes have different operational and control objectives. The micro-grid is connected to the distribution network through a substation transformer in grid-connected mode. The distribution network, in this case, is called the host grid or host network. It is important to note that the micro-grid need not connect to the low voltage section (LV $\approx \leq 1$ kV), which is sometimes called the secondary distribution network. It can also connect to the medium voltage section (MV $\approx \leq 1$–69 kV), referred to as the primary distribution network (Hatziargyriou et al. 2007). In this mode, the distribution network acts as the slack bus. It can absorb the surplus power from the micro-grid or inject the power into the micro-grid during a shortage to maintain a power balance. It defines the frequency of the micro-grid and the voltage at the point of common coupling. The micro-grid should adhere to the grid frequency and not affect the PCC voltage. Generally, the DG units are operated at their MPP in grid-connected mode if their source is renewable. If it's not renewable, they operate as per the unit commitment schedules from the grid operator. Apart from active power, the micro-grid also contributes some reactive power to the grid if required or operates at unity power factor with the grid (Olivares et al. 2014). Load or generation shedding within a micro-grid is also an option if the net import/export power has hard limits based on operational strategies or contractual obligations.

The islanded mode can be an intentional one or an unintentional one. The intentional ones are mainly related to maintenance and power quality issues in the host grid. The unintentional ones are due to faults and other unforeseen events in the host grid or micro-grid. Also, some remote micro-grids are economically or technically infeasible to operate in the grid-connected mode (Olivares et al. 2014). Such micro-grids always operate in the islanded mode. In islanded mode, the micro-grid should meet the supply–demand balance by itself either by controlling the generation or load shedding. But in any case, it has to meet the critical load. Apart from load-supply balance, the micro-grid should be able to control the frequency and voltages at different buses and achieve high power quality standards (Olivares et al. 2014).

1.10 Key Takeaways

- In a conventional power system, power is generated in bulk at generating stations, transmitted to long distances at high voltages through transmission lines and then distributed to customers at low voltage levels at the distribution system.
- Shortage of fossil fuels, adverse climatic impacts, high transmission losses, heavy costs in expansion, and low energy efficiencies are the major drawbacks of the conventional power system setup.
- To overcome the above mentioned challenges in the conventional power systems, novel technologies are being introduced in all the three aspects of power system, i.e. generation, transmission, and distribution. The concept of distributed generation is one such novel technology on the generation side.
- The basic idea of a distributed generator (DG) is to utilize the local DERs which are closer to the consumers, i.e. at the distribution system sites, to generate a small amount of power to meet the local needs and inject surplus power into the utility system.
- Design and control of power conversion technologies with adherence to grid codes, less inertia, source intermittency and congestion of the distribution system are the major hurdles in the operation of a single DG or multiple DGs operating independent of each other in a grid-connected mode.
- The basic idea of micro-grid is to aggregate different DGs, loads both controllable and uncontrollable, and storage elements into an independent network that can be operated autonomously and connected to the grid's distribution network.
- The concept of micro-grid allows the DGs to participate in the frequency regulation and voltage control.
- From the utility point of view, a micro-grid acts as a single good citizen that obeys the grid codes.
- From the local customer point of view, it can be viewed as a reliable generation and distribution network that can be trusted for uninterrupted power supply irrespective of grid condition, at low prices, and high power quality.

References

Dondi P, Bayoumi D, Haederli C, Julian D, Suter M (2002) Network integration of distributed power generation. J Power Sources 106(1–2):1–9

Hatziargyriou N, Asano H, Iravani R, Marnay C (2007) Microgrids. IEEE Power Energy Mag 5(4):78–94

Katiraei F, Iravani R, Hatziargyriou N, Dimeas A (2008) Microgrids management. IEEE Power Energ Mag 6(3):54–65

Kroposki B, Lasseter R, Ise T, Morozumi S, Papathanassiou S, Hatziargyriou N (2008) Making microgrids work. IEEE Power Energ Mag 6(3):40–53

Lasseter B (2001) Microgrids (distributed power generation). In: 2001 IEEE power engineering society winter meeting. Conference proceedings (Cat. No. 01CH37194), vol 1. IEEE, pp 146–149

Lasseter RH (2002) Microgrids. In: 2002 IEEE power engineering society winter meeting. Conference proceedings (Cat. No. 02CH37309), vol 1. IEEE, pp 305–308

Lopes JP, Hatziargyriou N, Mutale J, Djapic P, Jenkins N (2007) Integrating distributed generation into electric power systems: a review of drivers, challenges and opportunities. Electric Power Syst Res 77(9):1189–1203

Olivares DE, Mehrizi-Sani A, Etemadi AH, Cañizares CA, Iravani R, Kazerani M, Hajimiragha AH, Gomis-Bellmunt O, Saeedifard M, Palma-Behnke R, Jiménez-Estévez GA (2014) Trends in microgrid control. IEEE Trans Smart Grid 5(4):1905–1919

Ramakumar R (2001) Role of distributed generation in reinforcing the critical electric power infrastructure. In: 2001 IEEE power engineering society winter meeting. Conference proceedings (Cat. No. 01CH37194), vol 1. IEEE, pp 139–139

Chapter 2
An Overview of Micro-grid Control

Abstract The chapter provides a detailed overview of micro-grid control. The control objectives of a control system in the micro-grid are different for different operational modes. Section 2.1 provides an elaborate explanation of the control objectives of the micro-grid in both grid-connected and islanded modes. Section 2.2 explains different control architectures that are possibly employed to achieve the desired objectives. The section explains the underlying principles, pros, and cons of centralized architecture, decentralized architecture, and hierarchical architecture. Section 2.3 is completely dedicated to hierarchical architecture. The section explains the different control levels involved in the hierarchical architecture, the objectives of each control level, and the methodologies that can be used to achieve these objectives.

Keywords Centralized architecture · Decentralized architecture · Hierarchical architecture · Primary control level · Secondary control level · Tertiary control level

2.1 Control Objectives in a Micro-grid

In simple terms, the control system of a micro-grid has three major objectives (Olivares et al. 2014):

- The demand–supply balance of active and reactive powers in the micro-grid
- Frequency and voltage control
- Smooth transition between the modes of micro-grid operation.

But it is not as simple as it looks. The control system has to undergo a lot of compulsions from the utility grid and the customers while achieving the above objectives. It has to look into various other parameters of the grid and micro-grid. The control system should make sure that the operating ranges and/or tolerance limits of these parameters should not be violated while achieving the above objectives. The parameters and their operational ranges are very specific to the mode of operation of the micro-grid. Hence it is important to understand what is expected from a micro-grid control system in a grid-connected mode and islanded mode separately.

P. Vidyasagar and K. Shanti Swarup, *Design and Development of Model Predictive Primary Control of Micro Grids*, Springer Tracts in Electrical and Electronics Engineering, https://doi.org/10.1007/978-981-19-5852-6_2

First, we look at the expectations of a control system in the grid-connected mode (Olivares et al. 2014; Fan et al. 2012). The frequency in a grid-connected mode is determined by the grid. The micro-grid can either follow the grid frequency without participating in the frequency control or can participate in the frequency control if it has sufficient capacity of power reserves for frequency regulation. In the case of voltages, the micro-grid control system should be able to maintain the rated voltages at generator buses and should keep the load bus voltages within limits. But it is very important to note that there are some bus voltages in the micro-grid that the control system should not affect. The grid will impose the voltages on these buses. The bus at the point of common coupling (PCC) and other buses which are in close proximity to the grid comes under this category. The micro-grid control system should not affect these bus voltages as they will impact the grid operation. Now let's come to the interesting part of active and reactive power generation. If the micro-grid is not participating in frequency control, then the active power extracted from renewable DGs generally corresponds to the maximum power point tracking (MPPT). On the other hand, the fossil fuel-based DGs and storages try to balance the micro-grid demand if it is not met by the renewable DGs. The fossil fuel DGs can also inject power into the grid during peak demand time, scheduled times of dispatch from unit commitment studies, and other critical times based on the instructions from the grid operators. If there is a shortage in the generation of the micro-grid and it is not able to meet the load, then the micro-grid control system should allow the grid to supply the power shortage. If the micro-grid is participating in the frequency control, then it should have some primary reserve provision for each DG unit, including renewable DGs. This corresponds to DG units operating at off MPPTs. Whenever there are disturbances in the frequency, these primary reserves will be utilized by the control system to make the micro-grid participate in the frequency control. The reactive power output from the DG units should make sure that the voltages at the generator buses are maintained at rated voltages and the voltages at load buses are within limits. But from the grid's perspective, the reactive power injected into the grid from the micro-grid can be determined from any one of the following three requirements:

- Power factor correction
- Constant reactive power injection into the grid
- Voltage control at the PCC if the micro-grid is allowed to participate in the voltage control.

The micro-grid control in islanded mode is more complex and challenging than the grid-connected mode (Olivares et al. 2014). The control system in the islanded mode is under severe compulsion to achieve critical demand–supply equilibrium under all circumstances. This is because of the fact that the utility grid is absent to manage any imbalance in the supply and demand. Hence a very accurate load-sharing mechanism among the DG units is absolutely necessary (Piagi and Lasseter 2006; Lopes et al. 2006). Also, the frequency and voltages are no longer imposed by the grid. Hence the control system should make sure that the power is injected from the DGs into the micro-grid buses at a specified frequency defined by the dominant generator, also sometimes called as grid forming unit or master unit. Coming to the voltages, the control system should ensure that the voltages are maintained at all buses in the micro-grid by supplying enough reactive power in the micro-grid (Olivares et al. 2014; Fan et al. 2012; Xiaofeng et al. 2011).

2.2 Control Architectures in a Micro-grid

A complete centralized control of micro-grids, as shown in Fig. 2.1, is the first architecture that was proposed. In a centralized architecture, all the decisions are taken at a single point by a centralized controller (control centre or simply central controller) (Olivares et al. 2014; Hatta and Kobayashi 2008). The decisions are then communicated to different DG units in the micro-grid. For making the decisions, a large amount of data is to be collected from various buses at every sampling instant, and then it is to be communicated to the central controller for processing. The collection of data, processing, and communication to and from the controller should happen within a very small amount of time. This is due to the fact that the sampling rates of the controller are very high in the micro-grid. The completely centralized control architecture is not a feasible architecture for the micro-grid control due to the following issues (Olivares et al. 2014):

- The processing of a huge amount of data in a small amount of time requires parallel computing.
- The collection of a large amount of data is another problem as it requires huge amounts of field-level data acquisition units.
- The most important issue is the extensive and effective communication requirements.
- A single point failure in communication or processing, or collection of data is enough for the complete collapse of the micro-grid functioning.

Due to the above issues, the micro-grid control turned from completely centralized to completely decentralized architecture, as shown in Fig. 2.2. In a decentralized control, every DG unit is provided with its own local controller. The local controller

Fig. 2.1 Completely centralized control architecture

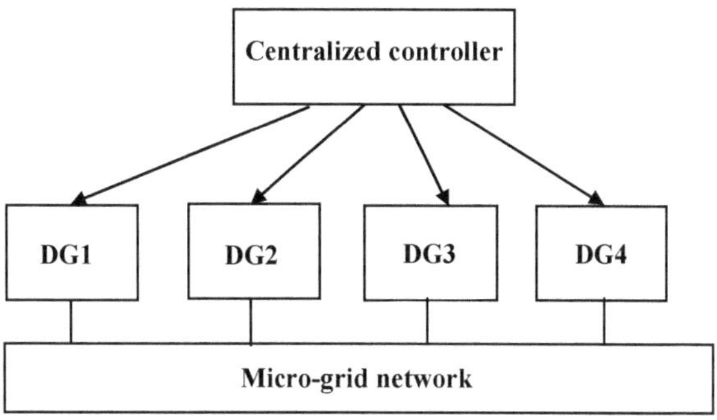

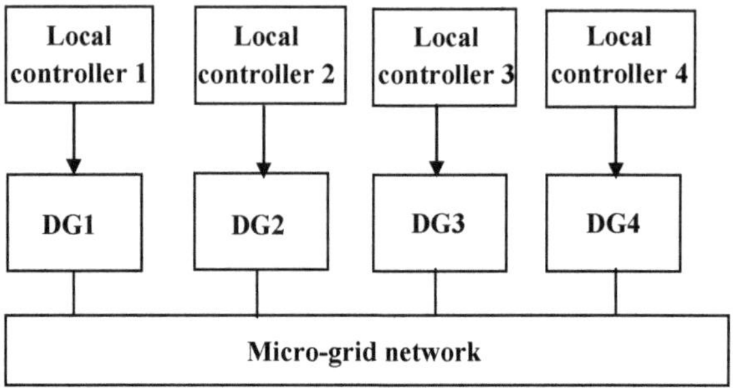

Fig. 2.2 Completely decentralized control architecture

gathers the information from its local bus and processes it in a pre-specified manner for generating the control decisions of the local DG unit. The controller sampling rate and processing speed in a completely decentralized architecture are very high. But there are some serious issues that need attention (Olivares et al. 2014; Ilić and Liu 1996; Ilic-Spong et al. 1988):

- The data processed by each controller is local to that bus. Hence the controllers are not aware of global system-wide information and other controller actions as there is no communication among them.
- The independent behaviour of the controllers leads to stability issues, non-optimal inefficient solutions, and highly oscillatory behaviour in the micro-grid.

Finally, we can come to a conclusion that both fully centralized and fully decentralized control architectures are infeasible in a micro-grid. The basic alternative then would be to look for a solution that is in between these two extremes. Such a solution already exists in conventional power systems. Due to their vast geographical areas, the interconnected power systems cannot implement fully centralized control systems as the communication requirements become infeasible. On the other hand, due to strong coupling between different power system parameters, a completely decentralized control leads to stability issues. Hence, the traditional power systems employ multi-layered control architecture called hierarchical architecture as a compromise between fully decentralized and centralized architectures, which require moderate communication and computational requirements, as shown in Fig. 2.3. The salient features of the multi-layered control architecture of a traditional power system are as follows (Marinovici et al. 2013):

- The entire control system is divided into different levels.
- The levels are vertically arranged, with each level having its own control objectives and sampling rate requirements.
- There will be a communication layer among different levels, which is not as complicated as in the case of fully centralized control.
- The response time of the lower levels is very low, and they perform the primary control actions in an immediate response to the disturbances in the system.
- The response time of the higher control levels is high, and they perform the optimal decision-making and corrective actions against the lower levels.

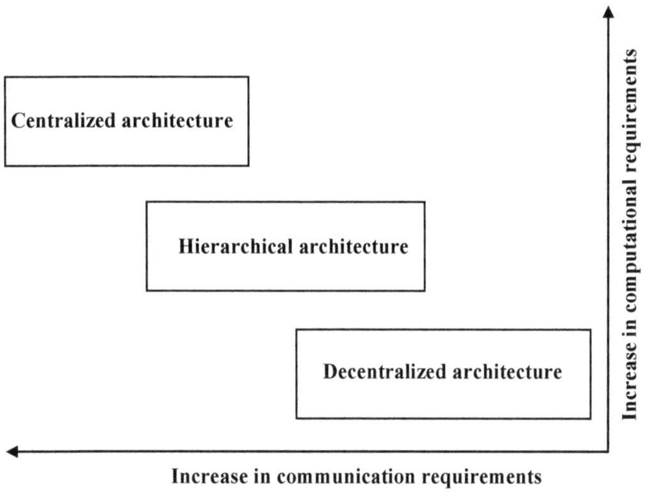

Fig. 2.3 Communication and computational requirements for different control architectures

- The lower levels implement a decentralized/centralized strategy based on the sampling rate requirements, whereas the higher levels are centralized in nature.

2.3 Hierarchical Control of a Standalone Micro-grid

A micro-grid experiences frequent topology changes, load disturbances, and source intermittencies. Under such a volatile environment, control architecture with reliable, robust, and adaptive features is expected. Large computational requirements, heavy data acquisition, and high speed communication of large amounts of data are discouraged in the control architecture. At the same time, priority to solve different tasks in the micro-grid should be given based on their time constants of operation. Fast-acting dynamics should be addressed immediately, followed by slow-acting dynamics. In view of these requirements, similar to multi-layered control architecture in traditional power systems, a hierarchical control structure is proposed for micro-grid as well. It is the modern way of performing micro-grid control (Olivares et al. 2014; Vasquez et al. 2010; Vandoorn et al. 2013a). It is one of the demanding research topics in power engineering. The hierarchical control structure consists of three control levels: primary control level, secondary control level, and tertiary control level, as shown in Fig. 2.4 (Vandoorn et al. 2013b). Each level had a different set of control tasks, different times of operation, and different communication requirements, as shown in Fig. 2.5. Apart from these differences, each control level employs different architecture, as shown in Fig. 2.6.

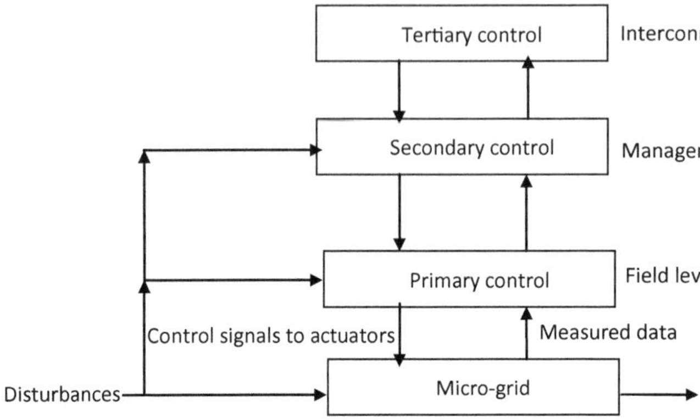

Fig. 2.4 Stages in hierarchical micro-grid control

2.3.1 Primary Control

The primary control level is the first level of control in the hierarchical architecture. It is also called the field level and has the fastest response among all the levels (Olivares et al. 2014; Andishgar et al. 2017). The typical response time is in 1 ms to a few hundred milliseconds range. It acts as the first line of defence for the micro-grid when there are disturbances. It may not bring the parameters to their nominal values and may not be economical and efficient. But it makes sure the micro-grid is in stable condition, and the micro-grid variables are in accepted ranges. The major responsibilities of the primary control level in an islanded mode are:

- Frequency control in the micro-grid
- Voltage control at different buses in the micro-grid
- Load sharing/power sharing among different DG units, so that demand–supply balance is maintained in the micro-grid.

The major responsibilities of the primary control level in the grid-connected mode are:

- Following the grid frequency at all times. If demanded by the grid operator, then it should participate in the frequency control as well in the gird-connected mode
- Voltage control at the buses that are not in the vicinity of the PCC. It should not disturb the grid-regulated voltage at the PCC. If demanded by the operator, then it can participate in the voltage control at the PCC as well.
- Make sure that the load-demand balance is achieved in the micro-grid. If there is surplus power, then it can inject the power into the grid. If there is a shortage, then it can absorb power from the grid
- Islanding detection
- Reactive power injection at PCC either to achieve:
  - Unity power factor at PCC

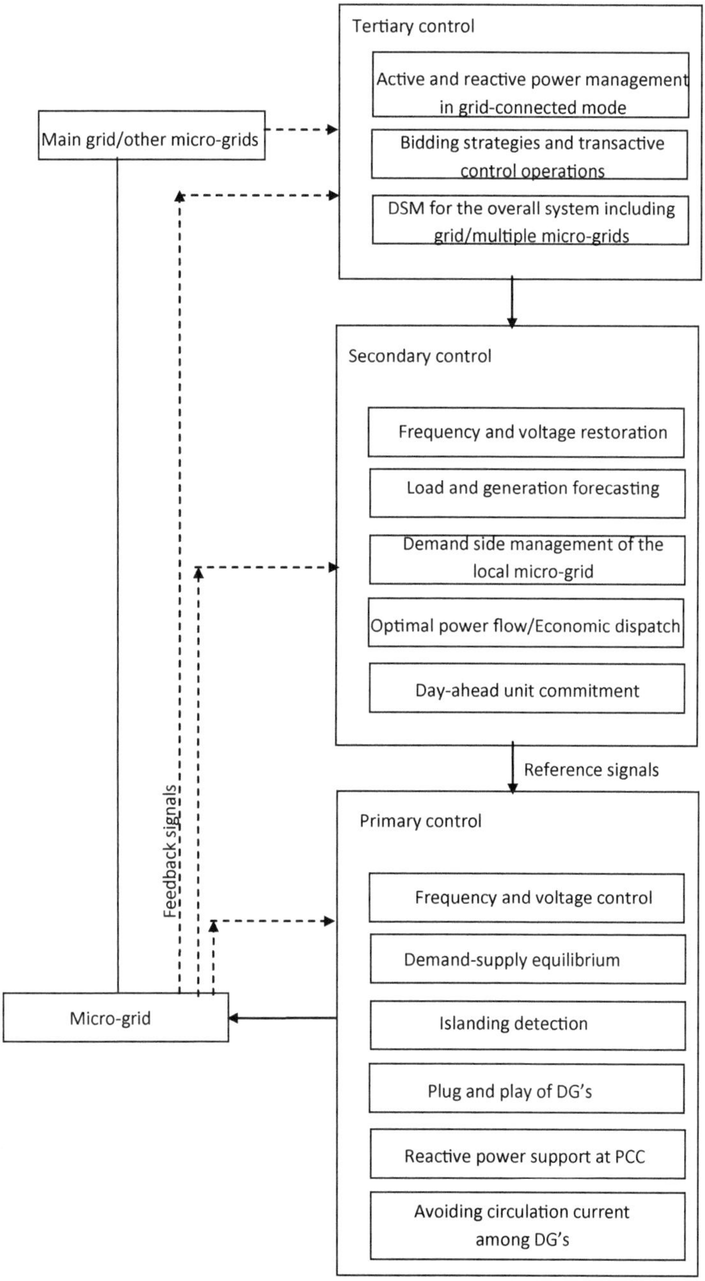

Fig. 2.5 Functions of different control levels in hierarchical micro-grid control

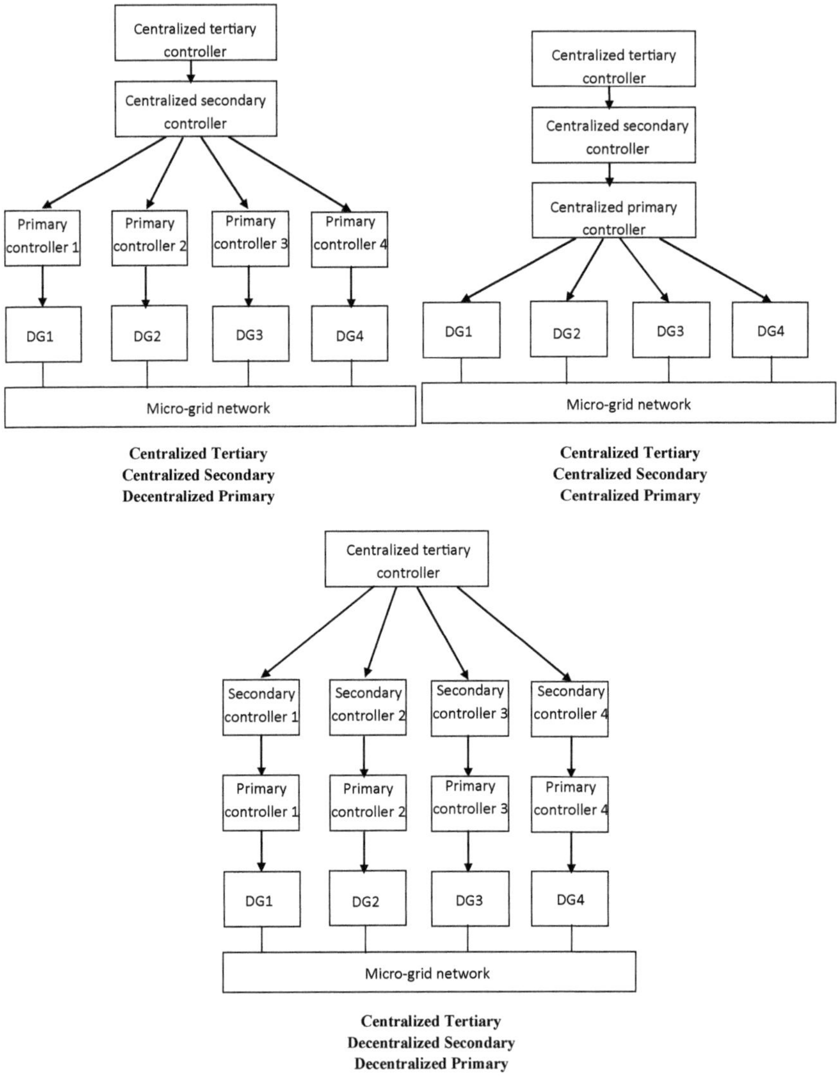

Fig. 2.6 Architectures of different control levels in hierarchical micro-grid control

- Voltage regulation at PCC
- Reactive power demanded by the grid operator.

Out of all the three control levels, the primary control level is the most challenging one for the control engineers. Speed is the major factor that influences the control architecture at the primary control level, as the sampling time lies in milliseconds. A high speed reliable communication and fast processing of data in a pre-specified manner to generate control signals are the major considerations while designing the primary control level architecture. Reliability, economy, and efficiency are out of bounds for the primary control level.

Conventional literatures on primary control divided it into two stages: power-sharing control and output control. The first stage of the primary control is the pre-specified power-sharing control strategy, where the load sharing of each DG unit for a particular load demand is decided. The power sharing is done either through communication-based control strategies or non-communication-based decentralized control strategies. In communication-based power sharing, the corresponding reference set-points are communicated to the local output controllers of each DG unit. Concentrated control, master/slave control, instantaneous current-sharing control, peak-value-based current-sharing control, centralized load-sharing control, current limitation control, circulating chain control, distributed control, and angle droop control are some of the important communication-based power-sharing strategies addressed in the literature (Vandoorn et al. 2013c). All these methods require system information (currents and voltages) gathered at a central location for processing and calculation of the references. These methods provide fast load sharing and good transient response. However, they are limited to micro-grid networks with concentrated load and generation. They are not suitable for micro-grids in which both generation and load are distributed throughout the micro-grid. Also, they are highly prone to communication failures (Olivares et al. 2014; Vandoorn et al. 2013c; Han et al. 2015).

Droop-based control is the widely used non-communication-based decentralized control strategy for power sharing (Olivares et al. 2014; Andishgar et al. 2017; Vandoorn et al. 2013c; Han et al. 2015; Sen and Kumar 2018; Guerrero et al. 2010). Each synchronous DG is provided with a speed governor that has an inherent frequency versus active power droop (f-P). The speed governor and inertia of the machine together are called the generation control system in the synchronous DG. This generation control system serves as the decentralized control strategy for power sharing in synchronous DGs. Each inverter-DG unit is equipped with its own power-sharing controller, which sets the reference points for its output controllers based on different frequency and voltage droop methods. There is a vast literature on the existing droop methods mentioning their pros and cons. However, droop-based decentralized control has several disadvantages like poor transient performance in networks with significant active and reactive power coupling, stability issues at different operating conditions due to improper droop coefficients, inability to impose constant nominal frequency independent of load, and failure or poor performance in the case of large load disturbances due to negligence of load dynamics.

Output control is the second stage of the primary control which is responsible for tracking the output voltage and current references of each DG unit set directly or indirectly by the power-sharing control. Decentralized strategy is by far the most used architecture for output control. Each synchronous DG is provided with the excitation control system, also called the automatic voltage regulator (AVR), for the output control stage. Each inverter-DG unit is equipped with an inner current control loop and outer voltage control loop, whereas each synchronous-DG unit is equipped with a generation control system and excitation control system (Olivares et al. 2014; Yu et al. 2015). PI, PID, PR, dead-beat, and hysteresis controllers are the most commonly used traditional controllers in the output control stage (Rocabert et al. 2012; Eid et al. 2014; Rokrok et al. 2018). The choice of the controller depends on the reference frame of analysis. Synchronous (*dq*) reference frame is associated with PI controllers, whereas stationary ($\alpha\beta$) reference frame with PR controllers. The natural (*abc*) reference frame is associated with any one of the PI, PR, dead-beat, or hysteresis controllers. These traditional output controllers of either synchronous DGs or inverter DGs require continuous parameter tuning to achieve stability and power quality. Designing the parameters of these output controllers is a herculean task as the micro-grid undergoes different operational changes and involves different operating points at different times. Several adaptive techniques, robust techniques, and intelligent techniques are proposed in the literature to design the parameters of these controllers (Mahmoud et al. 2017). Apart from the continuous tuning, these traditional controllers are not optimal and cannot handle the nonlinearity and system constraints at all times.

2.3.2 Secondary Control

The secondary control level is also called the management level. It is responsible for the reliable, secure, and economic operation of the micro-grids. It performs various energy management operations in the micro-grid and is hence also called an Energy Management System (EMS) (Olivares et al. 2014; Bidram and Davoudi 2012). Its response time is in a few minutes range. It is computationally more challenging than the primary control level. However, there is more time span between consecutive dispatch commands compared to the primary control level. In short, it can be viewed as an optimizer that performs a complex optimization problem in each sample. The optimal problem is nonlinear and non-convex. Some secondary controllers approximate the optimal control problem to a quadratic problem and some to a linear problem. The control problem can be single-objective or multi-objective at various instants in a day. The major objectives of the secondary control level are:

- Optimal day-ahead unit commitment studies
- Real-time economic dispatch/optimal power flow studies
- Restoration of frequency to its nominal value from the offset created by the primary control level
- Restoration of generator voltages to their nominal values
- Load forecasting
- Renewable source forecasting
- Incorporating demand-side management (DSM) strategies in the unit commitment and optimal power flow studies
- Creating bidding strategies for generators and loads in the micro-grid.

Secondary control is the highest control level in the standalone micro-grid. The secondary controller outputs form the set-points for the primary control level. The time frame of operation of the secondary control level is very slow compared to the primary level because of the following reasons:

- To decouple the operation of the primary and secondary control levels
- To allow the secondary controller to perform nonlinear optimization algorithms which are computationally very complex
- To reduce the high bandwidth communication, as the data that needs to be gathered and communicated back is very large compared to the primary control.

Coming to the methodologies, again, they are divided into centralized and decentralized methodologies. In a centralized secondary controller, the input information is gathered from different points in the micro-grid and is communicated to a central controller located at a single point. The central controller then solves an energy management optimization problem using this data, and the decisions are made, which are communicated to primary controllers as set-points. Unlike the primary control level, centralized control strategies are widely preferred at the secondary control level. The centralized controller should be provided with the following input information:

- Mathematical models of the DG units, storages, loads, and network
- Constant parameters of the network, DG units, storages, and loads
- Measurements and state estimation information about various dynamic parameters of the network, DG units, and loads
- Modes of operation of the micro-grid and the relevant information
- Various operational, security, and reliability constraints/limitations in the micro-grid
- Forecasted power capabilities of the renewable source DG units
- Forecasted micro-grid load data
- Cost functions involved in the optimization problem
- Parameter settings for the optimizer
- Forecasted electricity prices of the grid
- Storage SOC information
- DG unit offers that forms the generation bidding curves
- Load bids that form the load bidding curves.

Using the above input information, the secondary controller solves an optimization problem and produces the following output information:

- Unit commitment decision variables like DG unit on/off status
- DG unit economic dispatch decision variables like active power outputs and reactive power outputs/terminal voltages that forms the set-points for the primary controllers
- DSM-related outputs like load shutting and load shifting.

Now let us come to what kind of algorithms the secondary controller uses. In a very small micro-grid with very few possible scenarios of load and generation, the offline calculations of the optimal dispatch commands and storing them in look-up tables are very cost-effective and efficient in terms of performance. The sampling rate of the secondary controller can be set at rapid rates leading to very effective system performance. However, the look-up table method based on offline calculations is not a good idea if:

- The number of possible load and generation scenarios is very high
- The strong storage presence in the micro-grid which increases the scenario count
- Frequent changes in the topology of the micro-grid due to the faults or addition of the extra constituents.

To overcome the above issues with the look-up tables, the initial literature focussed on embedding the artificial neural networks (ANN) into the controller, which can train and learn different possible scenarios of operations of the micro-grid. But later, this method seems to be ineffective as the number of possible scenarios keeps on increasing in micro-grids, and the number of training sets is becoming more and more. Hence an online optimizer for solving the secondary controller optimal control problem at each sample is the modern methodology for the secondary control of micro-grids. As discussed above, the optimal energy management problem of a micro-grid is a non-convex nonlinear programming problem. Since unit commitment is an integral part of it, it can be called a mixed-integer nonlinear programming problem. The cost functions involve quadratic or higher-order polynomials, with some of the decision variables being integers. Apart from the complexity of cost functions, we also have very complex constraints on generation capacities, storage capacities, storage SOC limits, committed loads, controllable loads, network power flow constraints, and emission constraints. Sometimes, the objective function itself needs to involve the minimization of greenhouse gases along with the minimization of generation cost. Since these two are different cost functions, we need to solve a multi-objective control problem. Also, sometimes it will be necessary to involve generators and loads to participate in the bidding of their production and consumption at the secondary control level. Considering all these aspects, it is evident that both from the objective function perspective and constraints perspective, the optimal energy management problem at the secondary control level is a very hard problem to solve. To handle these kinds of problems, the micro-grid literature suggested heuristic optimization techniques like Genetic Algorithms (GA), PSO, and Ant Colony Optimization (ACO). Apart from these methodologies, some literature proposed adaptive controllers based on fuzzy controllers and ANN. Finally, an optimal controller based on the prediction of the future behaviour using the model of the micro-grid called model predictive controller (MPC) is the recently used advanced controller. The methodology of the MPC is the core of this book which will be discussed in detail in the subsequent chapters. But at this stage, it is worth noting that centralized MPC at the secondary control level of a micro-grid is massively successful.

So far, we have discussed the centralized secondary control strategies. But some literatures proposed decentralized strategies as well. The decentralized approach allows the interaction of the various units within the micro-grid in order to facilitate a distributed decision-making process. They can easily incorporate new DER units without the need to make continuous changes to the controller settings. Still, it has difficulties handling micro-grid operations requiring high levels of coordination.

Model predictive control (MPC) and multi-agent systems (MAS) are the widely proposed decentralized control strategies out of which MAS is more relatable than the MPC when it comes to decentralized architecture.

2.3.3 Tertiary Control

Tertiary control is the highest level of control in a micro-grid and is the interconnection level as it performs control actions related to the imports/exports of the micro-grid with the external grid or other micro-grid connections (Olivares et al. 2014; Bidram and Davoudi 2012). It exists only in the grid-connected mode or in a multi-micro-grid scenario. It is absent in an islanded micro-grid. It generates optimal set-points for the secondary control level based on the host grid requirements and data gathered from multiple micro-grids. Its operating time is higher than that of the secondary control and typically several minutes. Sometimes this control is considered as a part of the host grid and not the micro-grid. Similar to the secondary controller, the tertiary controller also solves an optimal energy management problem considering the technological, environmental, and economic criteria from the external environment and the micro-grid's dynamics. But the difference is that it solves the energy management problem for all the micro-grids and host grid combined together. The data measurement and communication it requires are much higher than that of the secondary control level. The major objectives of the tertiary control level are as follows:

- To determine whether micro-grids need to participate in frequency and voltage control of the overall system
- Reactive power management of the overall system
- DSM strategies over the entire system.

All the methodologies discussed above for the secondary control level can also work at the tertiary level. The only differences are:

- Data and communication requirements are high.
- The optimal energy management problem of a tertiary level requires more time to solve than that of the secondary level.

2.4 Key Takeaways

- Due to the absence of a host grid in standalone mode, the control system is continuously under the compulsion to achieve demand–supply equilibrium under all circumstances by implementing a proper load-sharing mechanism, frequency control, and voltage control within the micro-grid.
- In grid-connected mode, the control system should achieve critical demand–supply equilibrium within the micro-grid either through injecting surplus power

into grid or taking shortages from the grid. Frequency and voltage control are of least importance as they are mainly defined by the host grid.
- Centralized control architecture of a micro-grid consists of a central control unit that gathers the required information and processes it to determine the set-points for each of the distributed generator (DG) units from a single location.
- The centralized architecture becomes infeasible for micro-grids covering large geographical areas and having a large number of constituents (generators, loads, and feeders).
- In a decentralized architecture, every DG unit is provided with a local controller, which gathers the local bus information and processes it in a pre-specified manner to generate the local set-points for the DG.
- Lack of system awareness and a minimum level of coordination among DGs while taking the control decisions are the drawbacks of the decentralized architecture.
- A hierarchical control structure consists of three control levels, with each level having a different set of control tasks, different times of operation, and different architectural requirements.
- Hierarchical control is the modern way of performing micro-grid control and is one of the demanding research topics in power engineering.

References

Andishgar MH, Gholipour E, Hooshmand RA (2017) An overview of control approaches of inverter-based microgrids in islanding mode of operation. Renew Sustain Energy Rev 80:1043–1060

Bidram A, Davoudi A (2012) Hierarchical structure of microgrids control system. IEEE Trans Smart Grid 3(4):1963–1976

Eid BM, Abd Rahim N, Selvaraj J, El Khateb AH (2014) Control methods and objectives for electronically coupled distributed energy resources in microgrids: a review. IEEE Syst J 10(2):446–458

Fan Z, Zhang H, Guo L (2012) Simulation operation of inverters in microgrid under the island and grid. In: Proceedings of the 7th international power electronics and motion control conference, vol 3. IEEE, pp 2081–2084

Guerrero JM, Vasquez JC, Matas J, De Vicuña LG, Castilla M (2010) Hierarchical control of droop-controlled AC and DC microgrids—a general approach toward standardization. IEEE Trans Industr Electron 58(1):158–172

Han H, Hou X, Yang J, Wu J, Su M, Guerrero JM (2015) Review of power sharing control strategies for islanding operation of AC microgrids. IEEE Trans Smart Grid 7(1):200–215

Hatta H, Kobayashi H (2008) Demonstration study on centralized voltage control system for distribution line with sudden voltage fluctuations

Ilić MD, Liu SX (1996) Hierarchical power systems control: its value in a changing industry (advances in industrial control). Springer, London

Ilic-Spong M, Christensen J, Eichorn KL (1988) Secondary voltage control using pilot point information. IEEE Trans Power Syst 3(2):660–668

Lopes JP, Moreira CL, Madureira AG (2006) Defining control strategies for microgrids islanded operation. IEEE Trans Power Syst 21(2):916–924

Mahmoud MS, Alyazidi NM, Abouheaf MI (2017) Adaptive intelligent techniques for microgrid control systems: A survey. Int J Electr Power Energy Syst 90:292–305

Marinovici LD, Lian J, Kalsi K, Du P, Elizondo M (2013) Distributed hierarchical control architecture for transient dynamics improvement in power systems. IEEE Trans Power Syst 28(3):3065–3074

Olivares DE, Mehrizi-Sani A, Etemadi AH, Cañizares CA, Iravani R, Kazerani M, Hajimiragha AH, Gomis-Bellmunt O, Saeedifard M, Palma-Behnke R, Jiménez-Estévez GA (2014) Trends in microgrid control. IEEE Trans Smart Grid 5(4):1905–1919

Piagi P, Lasseter RH (2006) Autonomous control of microgrids. In: 2006 IEEE power engineering society general meeting. IEEE, 8-pp

Rocabert J, Luna A, Blaabjerg F, Rodriguez P (2012) Control of power converters in AC microgrids. IEEE Trans Power Electron 27(11):4734–4749

Rokrok E, Shafie-Khah M, Catalão JP (2018) Review of primary voltage and frequency control methods for inverter-based islanded microgrids with distributed generation. Renew Sust Energy Rev 82:3225–3235

Sen S, Kumar V (2018) Microgrid control: a comprehensive survey. Annu Rev Cont 45:118–151

Vandoorn TL, Vasquez JC, De Kooning J, Guerrero JM, Vandevelde L (2013a) Microgrids: hierarchical control and an overview of the control and reserve management strategies. IEEE Ind Electron Mag 7(4):42–55

Vandoorn TL, De Kooning JDM, Meersman B, Vandevelde L (2013c) Review of primary control strategies for islanded microgrids with power-electronic interfaces. Renew Sustain Energy Rev 19:613–628

Vandoorn TL, Guerrero JM, De Kooning JDM, Vásquez J, Vandevelde L (2013b) Decentralized and centralized control of islanded microgrids including reserve management. IEEE Ind Electron Mag 1–14

Vasquez JC, Guerrero JM, Miret J, Castilla M, De Vicuna LG (2010) Hierarchical control of intelligent microgrids. IEEE Ind Electron Mag 4(4):23–29

Xiaofeng S, Qingqiu L, Yanjun T, Chen Z (2011) An improved control method of power electronic converters in low voltage micro-grid. In: 2011 international conference on electrical machines and systems. IEEE, pp 1–6

Yu K, Ai Q, Wang S, Ni J, Lv T (2015) Analysis and optimization of droop controller for microgrid system based on small-signal dynamic model. IEEE Trans Smart Grid 7(2):695–705

Chapter 3
Mathematical Modelling of a Micro-grid

Abstract A generic modern-day micro-grid is of mixed nature. It should comprise both linear and nonlinear constituents in it. By this it means that the dynamical mathematical models of generators and loads should comprise both linear and nonlinear differential equations. It comprises rotating and non-rotating generators. This generally means presence of inertia in some generators like synchronous generators and induction generators, and the non-inertia generators like solar photovoltaic, battery energy storage systems, and fuel cell. It comprises rotating and non-rotating loads. This generally means presence of both static loads and rotating loads like induction motors and synchronous motors. It comprises both conventional synchronous generators and modern-day electronically interfaced generators like DGs with single inverter and DGs with back-to-back AC–DC–AC converters. It should contain buses that form or define the grid-connected mode, and a group of buses that can operate in the islanded mode in case grid-connected buses are disconnected from them. This chapter is all about detailed modelling of such a generic modern-day micro-grid which is the most important aspect for the design of the model predictive controller.

Keywords Electronically interfaced DG · Global reference frame · Local reference frame · Phase-Locked Loop · Synchronous-DG

3.1 Micro-grid Description and Reference Frames

A ten-bus micro-grid shown in Fig. 3.1 represents the topology of a modern-day generic micro-grid. This micro-grid test system is used throughout the book for the analysis of MPC and its MATLAB code formulations. It consists of all the basic constituents of a modern-day micro-grid described above, and hence, the analysis of this system can be extended to micro-grids with more number of constituents and covering more geographical area than this system. The test system consists of one synchronous-DG unit (SG-DG) and one electronically interfaced DG unit (EI-DG). The EI-DG can represent:

P. Vidyasagar and K. Shanti Swarup, *Design and Development of Model Predictive Primary Control of Micro Grids*, Springer Tracts in Electrical and Electronics Engineering, https://doi.org/10.1007/978-981-19-5852-6_3

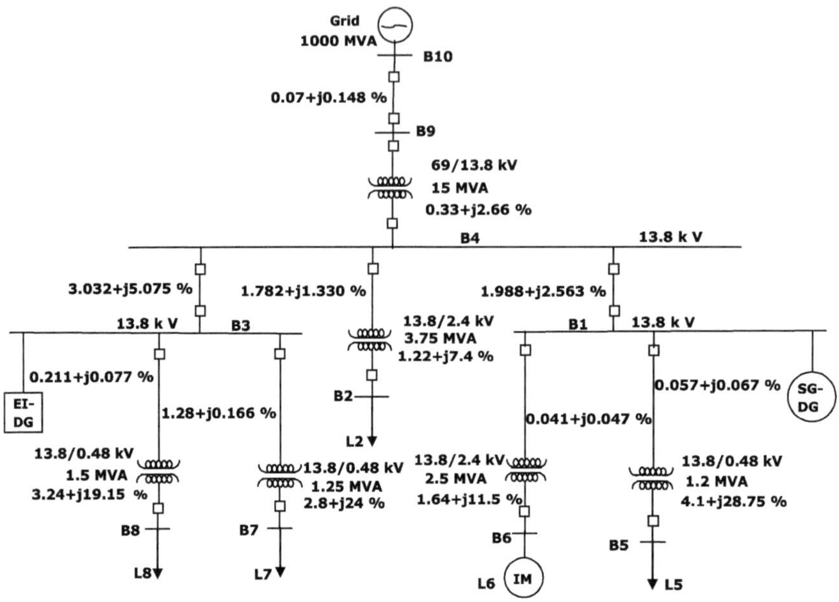

Fig. 3.1 Generic modern-day micro-grid

- Static generators like fuel cell, photovoltaic arrays, and battery energy storage system. These generators consist of a single inverter interfaced to the micro-grid.
- Rotating generators like synchronous generators supplied by micro-turbines and induction generators supplied by wind turbines interfaced to the micro-grid by a back-to-back AC–DC–AC converters.

In this book, the source of the EI-DG is considered to be photovoltaic array. Hence we have one conventional rotating generator (SG-DG) and one modern-day inverter-interfaced generator with a non-rotating source (EI-DG). Both the generators have highly nonlinear models. Micro-grid load (L2, L5, L6, L7, and L8) consists of impedance load and induction motor load. The load is aggregated at each load bus for the modelling purpose. The aggregated loads at buses B2, B5, B7, and B8 (L2, L5, L7, and L8) are purely impedance loads, and the aggregated load at bus B6 (L6) is purely induction motor load. The impedance loads represent linear and non-rotating loads. The induction motor load represents rotating and nonlinear load. It is evident from Fig. 3.1 that the micro-grid operates in grid-connected mode, if both B9 and B10 are present in the micro-grid with bus B4 connected to B9. The presence of both B9 and B10 defines the grid-connected mode of the micro-grid. Hence in grid-connected operation, the micro-grid can have 10 buses in it (B1–B10). If B4 is disconnected from B9, then the micro-grid becomes isolated and operates in standalone mode with the 8 buses (B1-B8). The per unit (p.u) impedance of the lines and transformers along with the voltage and MVA ratings of the transformers are shown in Fig. 3.1. The base values for the p.u calculations are defined on bus B1 with a base MVA of $MVA_b =$

5 MVA and a base kV of $kV_b = 13.8$ kV (line-line rms value). Details about the SG-DG parameters, EI-DG parameters, load parameters, and network parameters can be found in Appendix 1 at the end of the book.

Micro-grid analysis in the natural *abc* reference frame is a very difficult task. Two important reasons are the time-varying nature of the *abc* reference frame quantities and the model dependency on the rotor angles of the rotating machine, which are continuously changing with time. Hence, the micro-grid analysis is generally carried out in the rotating reference frames. Any component in the micro-grid is first modelled in the *abc* reference frame and then converted to its own rotating reference frame using Park's transformation. Different components in the micro-grid which are modelled in their own rotating reference frames are then finally converted to a common rotating reference frame. The entire micro-grid analysis is carried out in this common rotating reference frame.

Figure 3.2 shows the reference frames for the micro-grid shown in Fig. 3.1. The global reference frame *D*-*Q* rotating at an angular speed of ω_b rad/s (base speed) is defined on bus B10 with *D*-axis aligned along the voltage space vector $\boldsymbol{V}^{G10}$ of bus B10. Hence in the grid-connected mode, the global reference frame is defined on the utility bus. Since the grid is assumed to be a very stiff source, the frequency is more or less constant and hence the angular speed of *D*-*Q* frame ω_b is equal to the synchronous speed. The utility bus acts as slack bus in grid-connected mode. However, in standalone mode, the *D*-*Q* frame is initially defined on bus B1 with *D*-axis aligned along the voltage space vector $\boldsymbol{V}^{G1}$ of bus B1. The reason is that the bus B1 can act as a slack bus in the islanded mode due to the presence of a dominant SG-DG. But during the transient periods, the angular speed of *D*-*Q* frame ω_b is assumed to be constant at synchronous speed although the bus B1 frequency varies and may not be synchronous. This makes the *D*-*Q* frame an imaginary reference frame always rotating at synchronous speed and its position is adjusted during the steady-state so that it locks to the voltage space vector $\boldsymbol{V}^{G1}$ of bus B1 during the steady-state. This is necessary for the analysis as there is no dominant bus in the micro-grid that can provide a constant speed reference frame unlike grid. $d1$-$q1$ is the local reference frame of the SG-DG rotating at an angular speed of ω_r rad/s with $q1$-axis locked to its rotor. It is referred to as the rotor reference frame of the SG-DG. $d3$-$q3$ is the local reference frame of the PV-DG defined on bus B3 and rotating at an angular speed of ω_3 rad/s with $d3$-axis aligned along with the voltage space vector $\boldsymbol{V}^{G3}$ of bus B3. δ_1 and δ_3 are the angles between the local reference frame and global reference frame of the SG-DG and PV-DG, respectively. δ_0 is the torque angle or power angle of the SG-DG.

Let $\boldsymbol{f}_{abc}$ be a vector of instantaneous values (referred as "components" in this book) of a quantity '*f*' (either voltage or current) in *abc* frame. Let $\boldsymbol{f}_{dq0}$ be the vector of instantaneous values (components) of the same quantity '*f*' (either voltage or current) in a particular *dq*0 frame (referred simply as *d*-*q* frame in this book). The conversion of components of '*f*' from *abc* frame to *dq*0 (*d*-*q*) frame is given by the following park's transformation (Kundur et al. 1994):

$$\boldsymbol{f}_{dq0} = \boldsymbol{T}_{abc-dq0}\boldsymbol{f}_{abc} \tag{3.1}$$

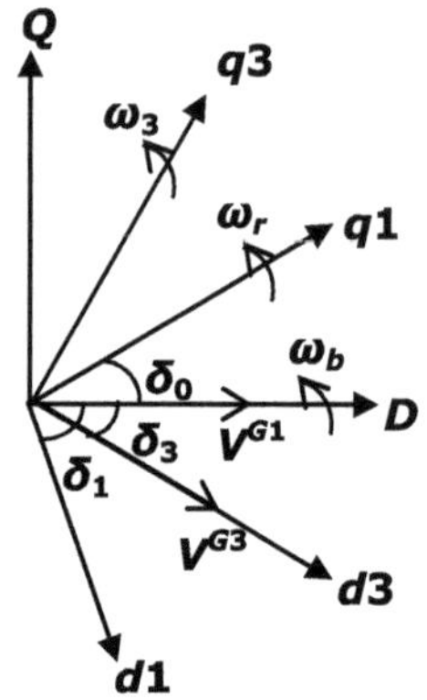

Fig. 3.2 Reference frames of the micro-grid

$$\boldsymbol{f}_{abc} = \begin{bmatrix} f_a \\ f_b \\ f_c \end{bmatrix}, \boldsymbol{f}_{dq0} = \begin{bmatrix} f_d \\ f_q \\ f_0 \end{bmatrix},$$

$$\boldsymbol{T}_{abc-dq0} = \frac{2}{3}\begin{bmatrix} \cos(\theta_c) & \cos\left(\theta_c - \frac{2\pi}{3}\right) & \cos\left(\theta_c + \frac{2\pi}{3}\right) \\ -\sin(\theta_c) & -\sin\left(\theta_c - \frac{2\pi}{3}\right) & -\sin\left(\theta_c + \frac{2\pi}{3}\right) \\ \frac{1}{2} & \frac{1}{2} & \frac{1}{2} \end{bmatrix}$$

The value of θ_c determines the local frame in which the components of 'f' are converted. For the conversion from *abc* frame to:

(i) *D*-*Q* frame: $\theta_c = \omega_b t$
(ii) $d1$-$q1$ frame: $\theta_c = \omega_b t + \delta_1$ and $\frac{d\delta_1}{dt} = \omega_r - \omega_b$
(iii) $d3$-$q3$ frame: $\theta_c = \omega_b t + \delta_3$ and $\frac{d\delta_3}{dt} = \omega_3 - \omega_b$.

3.2 Synchronous-DG Model

The SG-DG mathematical model comprises the dynamics of synchronous generator, turbine, and excitation system expressed in p.u. The synchronous generator dynamics and their mathematical formulations are extensively covered in the literature (Kundur et al. 1994; Krause et al. 2013). The synchronous generator dynamics are expressed in its rotor reference frame $d1$-$q1$. It mainly comprises the electrical and mechanical dynamics. In this book, the electrical dynamics of the synchronous generator are expressed by its subtransient model with stator dynamics included. The p.u electrical dynamics of the synchronous generator in its rotor reference frame $d1$-$q1$ are given by the following state vector:

$$X_{\textbf{SG-elec}} = \left[\, E'_{q1} \; E'_{d1} \; \Psi_{ad1} \; \Psi_{aq2} \; I_{d1} \; I_{q1} \,\right]^{\mathrm{T}}$$

where the corresponding dynamics are given by Kundur et al. (1994):

$$T'_{do}\frac{\mathrm{d}E'_{q1}}{\mathrm{d}t} = -E'_{q1} - \left(X_{d1} - X'_{d1}\right)\left(I_{d1} - \frac{\left(X'_{d1} - X''_{d1}\right)}{\left(X'_{d1} - X_{\mathrm{ls}}\right)^2}\right.$$
$$\left.\left(-E'_{q1} + \left(X'_{d1} - X_{\mathrm{ls}}\right)I_{d1} + \Psi_{ad1}\right)\right) + E_{fd} \tag{3.2}$$

$$T'_{qo}\frac{\mathrm{d}E'_{d1}}{\mathrm{d}t} = -E'_{d1} + \left(X_{q1} - X'_{q1}\right)\left(I_{q1} - \frac{\left(X'_{q1} - X''_{q1}\right)}{\left(X'_{q1} - X_{\mathrm{ls}}\right)^2}\right.$$
$$\left.\left(E'_{d1} + \left(X'_{q1} - X_{\mathrm{ls}}\right)I_{q1} + \Psi_{aq2}\right)\right) \tag{3.3}$$

$$T''_{do}\frac{\mathrm{d}\Psi_{ad1}}{\mathrm{d}t} = E'_{q1} - \left(X'_{d1} - X_{\mathrm{ls}}\right)I_{d1} - \Psi_{ad1} \tag{3.4}$$

$$T''_{qo}\frac{\mathrm{d}\Psi_{aq2}}{\mathrm{d}t} = -E'_{d1} - \left(X'_{q1} - X_{\mathrm{ls}}\right)I_{q1} - \Psi_{aq2} \tag{3.5}$$

$$\frac{1}{\omega_b}\frac{\mathrm{d}\Psi_{d1}}{\mathrm{d}t} = V_{d1} + R_s I_{d1} + \frac{\omega_r}{\omega_b}\Psi_{q1} \tag{3.6}$$

$$\frac{1}{\omega_b}\frac{\mathrm{d}\Psi_{q1}}{\mathrm{d}t} = V_{q1} + R_s I_{q1} - \frac{\omega_r}{\omega_b}\Psi_{d1} \tag{3.7}$$

$$\Psi_{d1} = -X''_{d1}I_{d1} + \frac{\left(X''_{d1} - X_{\mathrm{ls}}\right)}{\left(X'_{d1} - X_{\mathrm{ls}}\right)}E'_{q1} + \frac{\left(X'_{d1} - X''_{d1}\right)}{\left(X'_{d1} - X_{\mathrm{ls}}\right)}\Psi_{ad1} \tag{3.8}$$

$$\Psi_{q1} = -X''_{q1}I_{q1} - \frac{\left(X''_{q1} - X_{\mathrm{ls}}\right)}{\left(X'_{q1} - X_{\mathrm{ls}}\right)}E'_{d1} + \frac{\left(X'_{q1} - X''_{q}\right)}{\left(X'_{q1} - X_{\mathrm{ls}}\right)}\Psi_{aq2} \tag{3.9}$$

I_{d1}, I_{q1} are $d1$-$q1$ axis components of the SG-DG output current. V_{d1}, V_{q1} are $d1$-$q1$ axis components of the bus B1 voltage. E'_{d1}, E'_{q1} are $d1$-$q1$ axis components of the SG-DG stator induced voltage. Ψ_{ad1}, Ψ_{aq2} are $d1$-$q1$ axis components of the damper winding flux linkages. Ψ_{d1}, Ψ_{q1} are $d1$-$q1$ axis components of the armature flux linkages. T'_{do}, T'_{qo} are the transient open circuit time constants. T''_{do}, T''_{qo} are the subtransient open circuit time constants. R_s, X_{ls} are the stator resistance and leakage reactance. X''_{d1}, X''_{q1} are the subtransient reactance's. X'_{d1}, X'_{q1} are the transient reactance's. X_{d1}, X_{q1} are the synchronous reactance's. E_{fd} is the field voltage. The p.u mechanical dynamics of the SG-DG are given by:

$$X_{\textbf{SG-mech}} = \left[\,\delta_1\ \omega_r\,\right]^{\mathrm{T}}$$

where the corresponding dynamics are given by Kundur et al. (1994):

$$\frac{\mathrm{d}\delta_1}{\mathrm{d}t} = \omega_r - \omega_b \tag{3.10}$$

$$\frac{2H}{\omega_b}\frac{\mathrm{d}\omega_r}{\mathrm{d}t} = T_m - T_e \tag{3.11}$$

$$T_e = \Psi_{d1} I_{q1} - \Psi_{q1} I_{d1} \tag{3.12}$$

H is the inertia constant. T_e and T_m are electrical and mechanical torques exerted on the rotor. The turbine of the SG-DG is a single stage turbine which suits a conventional steam engine or a diesel engine. The p.u turbine dynamics are given by the following state vector:

$$X_{\mathbf{Turbine}} = \left[\, T_m \; P_{\mathrm{sv}} \,\right]^{\mathrm{T}}$$

where the corresponding dynamics are given by Kundur et al. (1994):

$$T_{\mathrm{CH}}\frac{\mathrm{d}T_m}{\mathrm{d}t} = -T_m + P_{\mathrm{sv}} \tag{3.13}$$

$$T_{\mathrm{sv}}\frac{\mathrm{d}P_{\mathrm{sv}}}{\mathrm{d}t} = -P_{\mathrm{sv}} + P_{\mathrm{ref}} - \frac{1}{R_D}\left(\frac{\omega_r}{\omega_b} - 1\right) \tag{3.14}$$

R_D is the droop constant of the governor. P_{ref} is the load reference set-point or turbine reference set-point. P_{sv} is the steam valve output. T_{sv} is the steam valve time constant. T_{CH} is the time constant of the steam chest. The p.u dynamic model of the excitation system is given by the following state vector:

$$X_{\mathbf{Excitation}} = \left[\, E_{fd} \; V_R \,\right]^{\mathrm{T}}$$

where the corresponding dynamics are given by Kundur et al. (1994):

$$T_E\frac{\mathrm{d}E_{fd}}{\mathrm{d}t} = V_R - K_E E_{fd} \tag{3.15}$$

$$T_A\frac{\mathrm{d}V_R}{\mathrm{d}t} = -V_R + K_A V_{\mathrm{in}} + K_A(1 - V_t) \tag{3.16}$$

V_t is the terminal voltage of the SG-DG, which is also the bus B1 voltage. K_E is the exciter gain, and T_E is its time constant. K_A is the amplifier gain, and T_A is its time constant. Voltage regulator's input is given by V_{in}, and its output is given by V_R.

3.3 EI-DG Model

The equivalent circuit of the EI-DG is shown in Fig. 3.3. As discussed, a PV source is considered in this book for the EI-DG. It is connected to micro-grid at bus B3 through a 3-Φ voltage source inverter (VSI). r_f, l_f, c_f are per phase resistance, inductance, and capacitance of the filter. r_t, l_t are per phase resistance and inductance of the interfacing transformer. C_{dc} is the DC link capacitor. V_{dc} is the DC link voltage across C_{dc}, and I_{cap} is the current through it. I_{pv} is the output current from the array, and I_{dc} is the DC link current. The vectors $\boldsymbol{e}_{3abc}$, $\boldsymbol{v}_{3fabc}$, and $\boldsymbol{v}_{3abc}$ consist of *abc* frame components (3-Φ instantaneous values) of VSI output voltage, filter capacitor voltage, and bus B3 voltage. Similarly, the vectors $\boldsymbol{i}_{3fabc}$ and $\boldsymbol{i}_{3abc}$ consist of *abc* frame components (3-Φ instantaneous values) of filter current and EI-DG output current. From the control perspective, both DC side and AC side dynamics of the VSI play a crucial role in determining the optimal control inputs to the EI-DG. Hence in this book, both DC and AC side dynamics of the VSI are considered for the analysis.

3.3.1 *AC Side Dynamics of the EI-DG in* abc *Frame*

The AC side dynamics of the EI-DG comprise filter dynamics and interfacing transformer dynamics. From Fig. 3.3, we can write the following AC side dynamics of the EI-DG:

$$\frac{\mathrm{d}\boldsymbol{i}_{3fabc}}{\mathrm{d}t} = \left(\frac{-r_f}{l_f}\right)\boldsymbol{i}_{3fabc} + \left(\frac{1}{l_f}\right)\left[\boldsymbol{e}_{3abc} - \boldsymbol{v}_{3fabc}\right] \tag{3.17}$$

$$\frac{\mathrm{d}\boldsymbol{i}_{3abc}}{\mathrm{d}t} = \left(\frac{-r_t}{l_t}\right)\boldsymbol{i}_{3abc} + \left(\frac{1}{l_t}\right)\left[\boldsymbol{v}_{3fabc} - \boldsymbol{v}_{3abc}\right] \tag{3.18}$$

$$\frac{\mathrm{d}\boldsymbol{v}_{3fabc}}{\mathrm{d}t} = \left(\frac{1}{c_f}\right)\left[\boldsymbol{i}_{3fabc} - \boldsymbol{i}_{3abc}\right] \tag{3.19}$$

$$\boldsymbol{i}_{3abc} = \begin{bmatrix} i_{3a} \\ i_{3b} \\ i_{3c} \end{bmatrix}, \boldsymbol{i}_{3fabc} = \begin{bmatrix} i_{3fa} \\ i_{3fb} \\ i_{3fc} \end{bmatrix}, \boldsymbol{v}_{3abc} = \begin{bmatrix} v_{3a} \\ v_{3b} \\ v_{3c} \end{bmatrix},$$

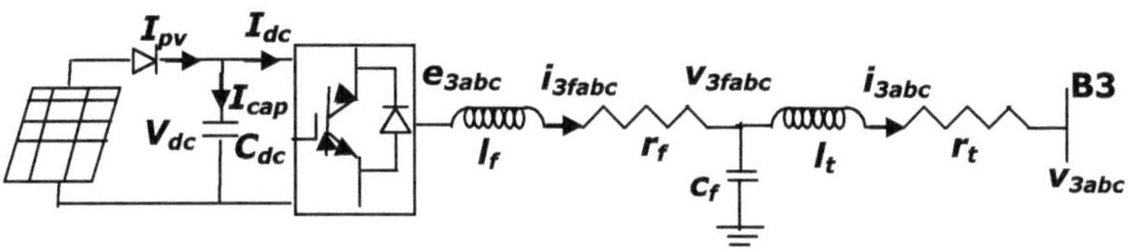

Fig. 3.3 EI-DG equivalent circuit

$$\boldsymbol{v}_{3fabc} = \begin{bmatrix} v_{3fa} \\ v_{3fb} \\ v_{3fc} \end{bmatrix}, \boldsymbol{e}_{3abc} = \begin{bmatrix} e_{3a} \\ e_{3b} \\ e_{3c} \end{bmatrix}$$

Let $\boldsymbol{E}_{3abc}$, $\boldsymbol{V}_{3fabc}$, $\boldsymbol{V}_{3abc}$, $\boldsymbol{I}_{3fabc}$, and $\boldsymbol{I}_{3abc}$ be the vectors that consist of p.u *abc* frame components of VSI output voltage, filter capacitor voltage, bus B3 voltage, filter current, and PV-DG output current. Let R_f, X_f, X_{Cf} be the p.u values of the filter parameters and R_t, X_t be the p.u values of the interfacing transformer. The AC side dynamics of the EI-DG given by (3.17)–(3.19) can be written in p.u as follows:

$$\frac{\mathrm{d}\boldsymbol{I}_{3fabc}}{\mathrm{d}t} = \left(\frac{-\omega_b R_f}{X_f}\right)\boldsymbol{I}_{3fabc} + \left(\frac{\omega_b}{X_f}\right)\left[\boldsymbol{E}_{3abc} - \boldsymbol{V}_{3fabc}\right] \tag{3.20}$$

$$\frac{\mathrm{d}\boldsymbol{I}_{3abc}}{\mathrm{d}t} = \left(\frac{-\omega_b R_t}{X_t}\right)\boldsymbol{I}_{3abc} + \left(\frac{\omega_b}{X_t}\right)\left[\boldsymbol{V}_{3fabc} - \boldsymbol{V}_{3abc}\right] \tag{3.21}$$

$$\frac{\mathrm{d}\boldsymbol{V}_{3fabc}}{\mathrm{d}t} = \left(\omega_b X_{cf}\right)\left[\boldsymbol{I}_{3fabc} - \boldsymbol{I}_{3abc}\right] \tag{3.22}$$

$$\boldsymbol{I}_{3abc} = \begin{bmatrix} I_{3a} \\ I_{3b} \\ I_{3c} \end{bmatrix}, \boldsymbol{I}_{3fabc} = \begin{bmatrix} I_{3fa} \\ I_{3fb} \\ I_{3fc} \end{bmatrix}, \boldsymbol{V}_{3abc} = \begin{bmatrix} V_{3a} \\ V_{3b} \\ V_{3c} \end{bmatrix},$$

$$\boldsymbol{V}_{3fabc} = \begin{bmatrix} V_{3fa} \\ V_{3fb} \\ V_{3fc} \end{bmatrix}, \boldsymbol{E}_{3abc} = \begin{bmatrix} E_{3a} \\ E_{3b} \\ E_{3c} \end{bmatrix},$$

$$\boldsymbol{I}_{3abc} = \left(\frac{1}{I_{\text{base}}}\right)\boldsymbol{i}_{3abc}, \boldsymbol{I}_{3fabc} = \left(\frac{1}{I_{\text{base}}}\right)\boldsymbol{i}_{3fabc},$$

$$\boldsymbol{V}_{3abc} = \left(\frac{1}{V_{\text{base}}}\right)\boldsymbol{v}_{3abc}, \boldsymbol{V}_{3fabc} = \left(\frac{1}{V_{\text{base}}}\right)\boldsymbol{v}_{3fabc},$$

$$\boldsymbol{E}_{3abc} = \left(\frac{1}{V_{\text{base}}}\right)\boldsymbol{e}_{3abc}, Z_{\text{base}} = \left(\frac{V_{\text{base}}}{I_{\text{base}}}\right), R_f = \left(\frac{r_f}{Z_{\text{base}}}\right),$$

$$X_f = \left(\frac{\omega_b l_f}{Z_{\text{base}}}\right), X_{cf} = \left(\frac{1}{Z_{\text{base}}\omega_b c_f}\right),$$

$$R_t = \left(\frac{r_t}{Z_{\text{base}}}\right), X_t = \left(\frac{\omega_b l_t}{Z_{\text{base}}}\right), V_{\text{base}} = \frac{\sqrt{2}\,\text{kV}_\text{b}}{\sqrt{3}}, I_{\text{base}} = \frac{\sqrt{2}\,\text{MVA}_\text{b}}{\sqrt{3}\,\text{kV}_\text{b}}$$

3.3.2 AC Side Dynamics of the EI-DG in Its Local d3-q3 Frame

The p.u AC side dynamics of the EI-DG in *abc* frame given by (3.20)–(3.22) can be converted to its local reference frame $d3$-$q3$ using the conversion procedure explained in the Sect. 3.1. The p.u AC side dynamics of the EI-DG in its local frame $d3$-$q3$ are given by the following state vector:

$$X_{\mathbf{PV-AC}} = \left[\, I_{fd3}\ I_{fq3}\ V_{fd3}\ V_{fq3}\ I_{d3}\ I_{q3} \,\right]^{\mathrm{T}}$$

where the corresponding dynamics are given by Schauder and Mehta (1993):

$$\frac{1}{\omega_b}\frac{\mathrm{d}I_{fd3}}{\mathrm{d}t} = \frac{-R_f}{X_f}I_{fd3} + \frac{\omega_3}{\omega_b}I_{fq3} + \frac{1}{X_f}\left[E_{d3} - V_{fd3}\right] \tag{3.23}$$

$$\frac{1}{\omega_b}\frac{\mathrm{d}I_{fq3}}{\mathrm{d}t} = \frac{-R_f}{X_f}I_{fq3} - \frac{\omega_3}{\omega_b}I_{fd3} + \frac{1}{X_f}\left[E_{q3} - V_{fq3}\right] \tag{3.24}$$

$$\frac{1}{\omega_b}\frac{\mathrm{d}V_{fd3}}{\mathrm{d}t} = \frac{\omega_3}{\omega_b}V_{fq3} + X_{Cf}\left[I_{fd3} - I_{d3}\right] \tag{3.25}$$

$$\frac{1}{\omega_b}\frac{\mathrm{d}V_{fq3}}{\mathrm{d}t} = -\frac{\omega_3}{\omega_b}V_{fd3} + X_{Cf}\left[I_{fq3} - I_{q3}\right] \tag{3.26}$$

$$\frac{1}{\omega_b}\frac{\mathrm{d}I_{d3}}{\mathrm{d}t} = \frac{-R_t}{X_t}I_{d3} + \frac{\omega_3}{\omega_b}I_{q3} + \frac{1}{X_t}\left[V_{fd3} - V_{d3}\right] \tag{3.27}$$

$$\frac{1}{\omega_b}\frac{\mathrm{d}I_{q3}}{\mathrm{d}t} = \frac{-R_t}{X_t}I_{q3} - \frac{\omega_3}{\omega_b}I_{d3} + \frac{1}{X_t}\left[V_{fq3} - V_{q3}\right] \tag{3.28}$$

E_{d3}, V_{fd3}, V_{d3}, I_{fd3}, and I_{d3} are $d3$-axis components, and E_{q3}, V_{fq3}, V_{q3}, I_{fq3}, and I_{q3} are $q3$-axis components of the VSI output voltage, filter capacitor voltage, bus B3 voltage, filter current, and EI-DG output current, respectively.

3.3.3 Phase-Locked Loop Dynamics

A phase-locked loop (PLL) is used to estimate the angular speed ω_3 of the EI-DG's local reference frame $d3$-$q3$ at bus B3. The estimation of ω_3 is necessary due to the following reasons:

- In grid-connected mode, the EI-DG bus frequency ω_3 at bus B3 is strongly determined by the grid frequency. It is more or less constant and is equal to the base

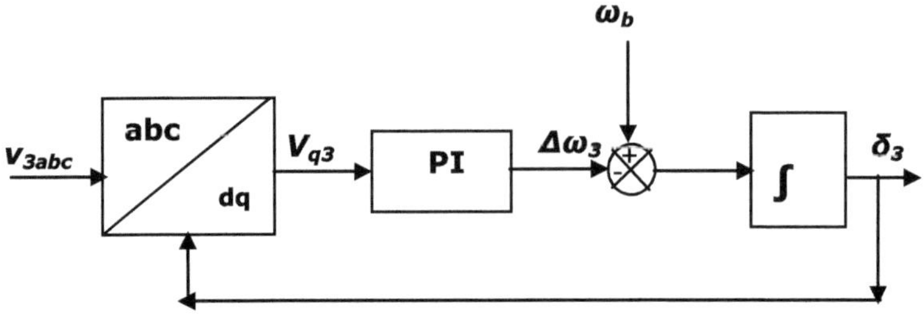

Fig. 3.4 PLL equivalent circuit

frequency ω_b determined by the grid. The PLL estimates this frequency ω_3 (basically same as ω_b), and the EI-DG simply follows this frequency (meaning it will inject active and reactive powers at this frequency).

- In islanded mode, the estimation of bus frequency ω_3 at bus B3 is necessary to implement an artificial droop characteristic for the EI-DG during the transient period to make it participate in the frequency regulation along with the SG-DG. It also makes sure that the EI-DG's output frequency is following the micro-grid's frequency which is determined by the rotor speed of the SG-DG in the case of islanded mode.

The equivalent circuit for PLL is shown in Fig. 3.4. Let K_p and K_i be the PLL parameters and φ_{PLL} be an intermediate state variable. Its p.u dynamics are given by:

$$X_{\mathbf{PLL}} = \left[\delta_3 \ \varphi_{\text{PLL}}\right]^{\text{T}}$$

where the corresponding dynamics are given by Chung (2000):

$$\frac{\mathrm{d}\delta_3}{\mathrm{d}t} = K_p V_{q3} + K_i \varphi_{\text{PLL}} \tag{3.29}$$

$$\frac{\mathrm{d}\varphi_{\text{PLL}}}{\mathrm{d}t} = V_{q3} \tag{3.30}$$

3.3.4 DC Side Dynamics of the EI-DG

The DC side model of the EI-DG can be obtained from the instantaneous power balance equation between the AC and DC sides of the EI-DG's inverter. The DC side model of the EI-DG can be represented by the following state vector:

$$X_{\mathbf{PV-DC}} = [V_{\text{dc}}]$$

where the corresponding dynamics are given by:

$$C_{dc}\frac{dV_{dc}}{dt} = I_{pv}(G, T) - \frac{(E_{d3}I_{fd3} + E_{q3}I_{fq3})\,MVA_b}{V_{dc}} \tag{3.31}$$

$I_{pv}(G, T)$ is the output current from the PV array which is a function of solar irradiance (G) and surface temperature (T) of the array (Chatterjee et al. 2011). The mathematical model of the PV array can be represented by the following algebraic equations:

$$\sigma = \frac{q}{AK(T + 273.15)N_{se}} \tag{3.32}$$

$$I_{sc}(T) = I_{scn} + K_I[T - T_n] \tag{3.33}$$

$$V_{oc}(T) = V_{ocn} + K_V[T - T_n] \tag{3.34}$$

$$I_0(T) = \left[\frac{I_{sc}(T)\left[1 + \frac{R_{se}}{R_{sh}}\right] - \frac{V_{oc}(T)}{R_{sh}}}{N_{pa}\left[e^{\sigma V_{oc}(T)} - e^{\sigma I_{sc}(T)R_{se}}\right]}\right] \tag{3.35}$$

$$I_{ph}(T) = I_0(T)\left[e^{\sigma V_{oc}(T)} - 1\right] + \frac{V_{oc}(T)}{R_{sh}N_{pa}} \tag{3.36}$$

$$I_{sc}(G, T) = [I_{scn} + K_I(T - T_n)]\frac{G}{G_n} \tag{3.37}$$

$$I_{ph}(G, T) = \left[I_0(T)\left[e^{\sigma V_{oc}(T)} - 1\right] + \frac{V_{oc}(T)}{R_{sh}N_{pa}}\right]\frac{G}{G_n} \tag{3.38}$$

$$I_0(G, T) = \left[\frac{N_{pa}I_{ph}(G, T) - I_{sc}(G, T)\left[1 + \frac{R_{se}}{R_{sh}}\right]}{N_{pa}\left[e^{\sigma I_{sc}(G,T)R_{se}} - 1\right]}\right] \tag{3.39}$$

$$V_{oc}(G, T) = R_{sh}N_{pa}I_{ph}(G, T) - R_{sh}N_{pa}I_0(G, T)\left[e^{\sigma V_{oc}(G,T)} - 1\right] \tag{3.40}$$

$$I_{pv}(G, T) = N_{pp}\left[N_{pa}I_{ph}(G, T) - N_{pa}I_0(G, T)\left(e^{\sigma\left[\frac{V_{dc}}{N_{ss}} + \frac{I_{pv}(G,T)R_{se}}{N_{pp}}\right]} - 1\right) - \frac{\frac{V_{dc}}{N_{ss}} + \frac{I_{pv}(G,T)R_{se}}{N_{pp}}}{R_{sh}}\right] \tag{3.41}$$

q is the electron charge. A is the diode quality factor. K is the Boltzmann's constant. N_{pa} is the number of parallel rows in each module. N_{se} is the number of series cells in each parallel row of a module. I_{sc} is the short circuit current of the module. I_{scn} is the short circuit current of the module at nominal conditions. K_I is the temperature coefficient of the short circuit current. T_n is the nominal temperature. V_{oc} is the open circuit voltage of the module. V_{ocn} is the open circuit voltage of the module at nominal conditions. K_V is the temperature coefficient of the open circuit voltage. I_0 is the dark saturation current of the diode. R_{se} is the equivalent series resistance of each module. R_{sh} is the equivalent shunt resistance of each module. I_{ph} is the photo-generated current of each solar cell. G_n is the nominal irradiance. N_{pp} is the number of parallel rows of modules in the PV array. N_{ss} is the number of series modules in each parallel row of the PV array.

3.4 Load Modelling in the Micro-grid

All the loads are modelled in global *D-Q* reference frame. The loads at buses B2, B5, B7, and B8 are impedance loads. The impedance loads at each of these buses are aggregated and represented as L2, L5, L7, and L8, respectively. The aggregate loads L2, L5, L7, and L8 are modelled as p.u R-L loads in *abc* reference frame. From *abc* reference frame, the models are converted into global *D-Q* reference frame using the procedure explained in Sect. 3.1. At any *i*th bus, the p.u dynamic model of an aggregated R-L load in global *D-Q* reference frame is given by the following state vector:

$$\boldsymbol{X}_{\mathbf{Load}i} = \left[\, I_{li}^{D} \; I_{li}^{Q} \,\right]^{\mathrm{T}}$$

where the corresponding dynamics are given by Katiraei et al. (2007):

$$\frac{1}{\omega_b}\frac{\mathrm{d}I_{li}^{D}}{\mathrm{d}t} = \frac{-R_{li}}{X_{li}} I_{li}^{D} + I_{li}^{Q} + \frac{1}{X_{li}}\left[V_i^{D}\right] \tag{3.42}$$

$$\frac{1}{\omega_b}\frac{\mathrm{d}I_{li}^{Q}}{\mathrm{d}t} = \frac{-R_{li}}{X_{li}} I_{li}^{Q} - I_{li}^{D} + \frac{1}{X_{li}}\left[V_i^{Q}\right] \tag{3.43}$$

V_i^D, V_i^Q are *D-Q* axis components of *i*th bus voltage. I_{li}^D, I_{li}^Q are *D-Q* axis components of the load current at *i*th bus. R_{li}, X_{li} are p.u resistance and reactance of *i*th bus load. The load at bus B6 is an aggregated induction motor load L6. At any *i*th bus, the p.u dynamics of the induction motor load in global *D-Q* reference frame are given by the following state vector:

$$\boldsymbol{X}_{\mathbf{IM}i} = \left[\, E_{mi}^{D} \; E_{mi}^{Q} \; I_{mi}^{D} \; I_{mi}^{Q} \; s_{mi} \,\right]^{\mathrm{T}}$$

where the corresponding fifth-order dynamics are given by Kundur et al. (1994):

$$T_{mi}\frac{dE_{mi}^{D}}{dt} = -\left[E_{mi}^{D} + \left(X_{mi} - X'_{mi}\right)I_{mi}^{Q}\right] + \omega_b s_{mi} T_{mi} E_{mi}^{Q} \tag{3.44}$$

$$T_{mi}\frac{dE_{mi}^{Q}}{dt} = -\left[E_{mi}^{Q} - \left(X_{mi} - X'_{mi}\right)I_{mi}^{D}\right] - \omega_b s_{mi} T_{mi} E_{mi}^{D} \tag{3.45}$$

$$T_{mi}\frac{dI_{mi}^{D}}{dt} = \frac{E_{mi}^{Q} + \omega_b(s_{mi}-1)T_{mi}E_{mi}^{D} + \omega_b T_{mi} V_i^{D}}{X'_{mi}} + \omega_b T_{mi} I_{mi}^{Q} - \frac{\left(\omega_b T_{mi} R_{mi} + \left(X_{mi} - X'_{mi}\right)\right)I_{mi}^{D}}{X'_{mi}} \tag{3.46}$$

$$T_{mi}\frac{dI_{mi}^{Q}}{dt} = \frac{-E_{mi}^{D} + \omega_b(s_{mi}-1)T_{mi}E_{mi}^{Q} + \omega_b T_{mi} V_i^{Q}}{X'_{mi}} - \omega_b T_{mi} I_{mi}^{D} - \frac{\left(\omega_b T_{mi} R_{mi} + \left(X_{mi} - X'_{mi}\right)\right)I_{mi}^{Q}}{X'_{mi}} \tag{3.47}$$

$$2H_{mi}\frac{ds_{mi}}{dt} = T_{mmi} - T_{emi} \tag{3.48}$$

I_{mi}^{D}, I_{mi}^{Q} are *D-Q* axis components of *i*th bus induction motor load current. E_{mi}^{D}, E_{mi}^{Q} are *D-Q* axis components of the induced voltage. R_{mi}, X'_{mi}, X_{mi} are stator resistance, transient reactance, and synchronous reactance. T_{mi} is the transient open circuit time constant. S_{mi} is the slip. H_{mi} is the inertia constant. T_{emi}, T_{mmi} are electrical and mechanical torques exerted on the rotor.

3.5 Network Modelling in the Micro-grid

The network model includes the R-L dynamics of the network lines and the shunt capacitor dynamics at each bus. The shunt capacitor at each bus comes from the Π model of the network lines connected to that bus and the passive capacitor banks that are installed at the bus to improve its voltage profile. All the network dynamics are expressed in global *D-Q* frame. The p.u dynamic model of any line connected between buses '*i*' and '*j*' in global *D-Q* reference frame is given by the following state vector:

$$X_{Lij} = \left[I_{ij}^{D}\ I_{ij}^{Q}\right]^{T}$$

where the corresponding dynamics are given by Katiraei et al. (2007):

$$\frac{1}{\omega_b}\frac{dI_{ij}^{D}}{dt} = \frac{-R_{ij}}{X_{ij}}I_{ij}^{D} + I_{ij}^{Q} + \frac{1}{X_{ij}}\left[V_i^{D} - V_j^{D}\right] \tag{3.49}$$

$$\frac{1}{\omega_b}\frac{\mathrm{d}I_{ij}^{Q}}{\mathrm{d}t} = \frac{-R_{ij}}{X_{ij}}I_{ij}^{Q} - I_{ij}^{D} + \frac{1}{X_{ij}}\left[V_i^Q - V_j^Q\right] \tag{3.50}$$

I_{ij}^D, I_{ij}^Q are *D-Q* axis components of the line current. R_{ij}, X_{ij} are p.u resistance and inductance of the line. Let the number of buses in the micro-grid be *NB*. At any *i*th bus, the p.u dynamics of the shunt capacitance in global *D-Q* reference frame are given by the following state vector:

$$\boldsymbol{X}_{\boldsymbol{L}\boldsymbol{i}} = \left[\, V_i^D \; V_i^Q \,\right]^{\mathrm{T}}$$

where the corresponding dynamics are given by Katiraei et al. (2007):

$$\frac{1}{\omega_b}\frac{\mathrm{d}V_i^D}{\mathrm{d}t} = V_i^Q + X_{Ci}\left(I_i^D - I_{li}^D - I_{mi}^D - \sum_{\substack{j=1\\ j\neq i}}^{\mathrm{NB}} I_{ij}^D \right) \tag{3.51}$$

$$\frac{1}{\omega_b}\frac{\mathrm{d}V_i^Q}{\mathrm{d}t} = -V_i^D + X_{Ci}\left(I_i^Q - I_{li}^Q - I_{mi}^Q - \sum_{\substack{j=1\\ j\neq i}}^{\mathrm{NB}} I_{ij}^Q \right) \tag{3.52}$$

I_i^D, I_i^Q are *D-Q* axis components of the generator injected current at *i*th bus. *NB* is the number of buses. X_{Ci} is the p.u shunt capacitance at *i*th bus.

3.6 Complete Model of the Grid-Connected Micro-grid

The complete grid-connected mathematical model of the micro-grid showed in Fig. 3.1 can be expressed as a nonlinear state-space model given by:

$$\frac{\mathrm{d}\boldsymbol{X}(t)}{\mathrm{d}t} = \boldsymbol{f}(\boldsymbol{X}(t), \boldsymbol{V}(t))$$

$$\boldsymbol{Y}(t) = \boldsymbol{g}(\boldsymbol{X}(t)) \tag{3.53}$$

X is the continuous-time state vector with *n* states, *V* is the input vector with n_{ip} inputs and *Y* is the output vector with n_{op} outputs given by:

$$\begin{aligned}\boldsymbol{X}^{\mathrm{T}} = \big[\, & \boldsymbol{X}_{\mathbf{SG-elec}}^{\mathrm{T}}\, \boldsymbol{X}_{\mathbf{SG-mech}}^{\mathrm{T}}\, \boldsymbol{X}_{\mathbf{Turbine}}^{\mathrm{T}}\, \boldsymbol{X}_{\mathbf{Exciatation}}^{\mathrm{T}}\, \boldsymbol{X}_{\mathbf{PV-AC}}^{\mathrm{T}}\, \boldsymbol{X}_{\mathbf{PLL}}^{\mathrm{T}}\, \boldsymbol{X}_{\mathbf{PV-DC}}^{\mathrm{T}} \\ & \boldsymbol{X}_{\mathbf{Load2}}^{\mathrm{T}}\, \boldsymbol{X}_{\mathbf{Load5}}^{\mathrm{T}}\, \boldsymbol{X}_{\mathbf{Load7}}^{\mathrm{T}}\, \boldsymbol{X}_{\mathbf{Load8}}^{\mathrm{T}}\, \boldsymbol{X}_{\mathbf{IM6}}^{\mathrm{T}}\, \boldsymbol{X}_{\mathbf{L15}}^{\mathrm{T}}\, \boldsymbol{X}_{\mathbf{L16}}^{\mathrm{T}}\, \boldsymbol{X}_{\mathbf{L14}}^{\mathrm{T}}\, \boldsymbol{X}_{\mathbf{L24}}^{\mathrm{T}}\, \boldsymbol{X}_{\mathbf{L34}}^{\mathrm{T}} \\ & \boldsymbol{X}_{\mathbf{L37}}^{\mathrm{T}}\, \boldsymbol{X}_{\mathbf{L38}}^{\mathrm{T}}\, \boldsymbol{X}_{\mathbf{L49}}^{\mathrm{T}}\, \boldsymbol{X}_{\mathbf{L910}}^{\mathrm{T}}\, \boldsymbol{X}_{\mathbf{L1}}^{\mathrm{T}}\, \boldsymbol{X}_{\mathbf{L2}}^{\mathrm{T}}\, \boldsymbol{X}_{\mathbf{L3}}^{\mathrm{T}}\, \boldsymbol{X}_{\mathbf{L4}}^{\mathrm{T}}\, \boldsymbol{X}_{\mathbf{L5}}^{\mathrm{T}}\, \boldsymbol{X}_{\mathbf{L6}}^{\mathrm{T}}\, \boldsymbol{X}_{\mathbf{L7}}^{\mathrm{T}}\, \boldsymbol{X}_{\mathbf{L8}}^{\mathrm{T}}\, \boldsymbol{X}_{\mathrm{L9}}^{\mathrm{T}} \,\big]\end{aligned}$$

$$\boldsymbol{V}^{\mathrm{T}} = \left[\, P_{\mathrm{ref}}\ V_{\mathrm{in}}\ E_{d3}\ E_{q3} \,\right]$$

$$\boldsymbol{Y}^{\mathrm{T}} = \left[\, P_{G1}\ V_1\ V_3\ P_{G3} \,\right] \rightarrow \text{voltage control mode}$$

$$\boldsymbol{Y}^{\mathrm{T}} = \left[\, P_{G1}\ Q_{G1}\ Q_{G3}\ P_{G3} \,\right] \rightarrow \text{power factor correction mode (or) constant power injection mode}$$

P_{G3} and Q_{G3} are the active power and reactive power injected at bus B3 by the EI-DG. P_{G1} and Q_{G1} are the active power and reactive power injected at bus B1 by the SG-DG. V_1 and V_3 are the voltages at generator buses B1 and B3. There are a total of $n = 70$ states, $n_{\mathrm{ip}} = 4$ inputs, and $n_{\mathrm{op}} = 4$ outputs in the grid-connected mode of the micro-grid shown in Fig. 3.1.

3.7 Complete Model of the Standalone Micro-grid

The complete standalone mathematical model of the micro-grid shown in Fig. 3.1 can be expressed as a nonlinear state-space model given by:

$$\frac{\mathrm{d}\boldsymbol{X}(t)}{\mathrm{d}t} = \boldsymbol{f}(\boldsymbol{X}(t), \boldsymbol{V}(t))$$

$$\boldsymbol{Y}(t) = \boldsymbol{g}(\boldsymbol{X}(t)) \tag{3.54}$$

X is the continuous-time state vector with n states, V is the input vector with n_{ip} inputs, and Y is the output vector with n_{op} outputs given by:

$$\begin{aligned}\boldsymbol{X}^{\mathrm{T}} = \big[\, & \boldsymbol{X}^{\mathrm{T}}_{\mathbf{SG-elec}}\ \boldsymbol{X}^{\mathrm{T}}_{\mathbf{SG-mech}}\ \boldsymbol{X}^{\mathrm{T}}_{\mathbf{Turbine}}\ \boldsymbol{X}^{\mathrm{T}}_{\mathbf{Exciatation}}\ \boldsymbol{X}^{\mathrm{T}}_{\mathbf{PV-AC}}\ \boldsymbol{X}^{\mathrm{T}}_{\mathbf{PLL}}\ \boldsymbol{X}^{\mathrm{T}}_{\mathbf{PV-DC}} \\ & \boldsymbol{X}^{\mathrm{T}}_{\mathbf{Load2}}\ \boldsymbol{X}^{\mathrm{T}}_{\mathbf{Load5}}\ \boldsymbol{X}^{\mathrm{T}}_{\mathbf{Load7}}\ \boldsymbol{X}^{\mathrm{T}}_{\mathbf{Load8}}\ \boldsymbol{X}^{\mathrm{T}}_{\mathbf{IM6}}\ \boldsymbol{X}^{\mathrm{T}}_{\mathbf{L15}}\ \boldsymbol{X}^{\mathrm{T}}_{\mathbf{L16}}\ \boldsymbol{X}^{\mathrm{T}}_{\mathbf{L14}}\ \boldsymbol{X}^{\mathrm{T}}_{\mathbf{L24}}\ \boldsymbol{X}^{\mathrm{T}}_{\mathbf{L34}} \\ & \boldsymbol{X}^{\mathrm{T}}_{\mathbf{L37}}\ \boldsymbol{X}^{\mathrm{T}}_{\mathbf{L38}}\ \boldsymbol{X}^{\mathrm{T}}_{\mathbf{L1}}\ \boldsymbol{X}^{\mathrm{T}}_{\mathbf{L2}}\ \boldsymbol{X}^{\mathrm{T}}_{\mathbf{L3}}\ \boldsymbol{X}^{\mathrm{T}}_{\mathbf{L4}}\ \boldsymbol{X}^{\mathrm{T}}_{\mathbf{L5}}\ \boldsymbol{X}^{\mathrm{T}}_{\mathbf{L6}}\ \boldsymbol{X}^{\mathrm{T}}_{\mathbf{L7}}\ \boldsymbol{X}^{\mathrm{T}}_{\mathbf{L8}} \,\big]\end{aligned}$$

$$\boldsymbol{V}^{\mathrm{T}} = \left[\, P_{\mathrm{ref}}\ V_{\mathrm{in}}\ E_{d3}\ E_{q3} \,\right]$$

$$\boldsymbol{Y}^{\mathrm{T}} = \left[\, \omega_3\ V_1\ V_3\ P_{G3} \,\right]$$

There are a total of $n = 64$ states, $n_{\mathrm{ip}} = 4$ inputs, and $n_{\mathrm{op}} = 4$ outputs in the standalone mode of the micro-grid shown in Fig. 3.1.

3.8 Key Takeaways

- The chapter discussed the detailed mathematical model of the generic modern-day micro-grid.
- Each and every component of the micro-grid, i.e., generators, lines, impedance loads, induction motor loads, and shunt capacitances of the buses are modelled to a very minute level.
- The complete mathematical model of the micro-grid is expressed as a nonlinear state-space model.
- This complete dynamic model of the micro-grid is highly nonlinear.
- The mathematical model discussed in this chapter serves as a fundamental building block for the MPC formulations discussed in the subsequent chapters.

References

Chatterjee A, Keyhani A, Kapoor D (2011) Identification of photovoltaic source models. IEEE Trans Energy Convers 26(3):883–889

Chung SK (2000) A phase tracking system for three phase utility interface inverters. IEEE Trans Power Electron 15(3):431–438

Katiraei F, Iravani MR, Lehn PW (2007) Small-signal dynamic model of a micro-grid including conventional and electronically interfaced distributed resources. IET Gener Trans Distrib 1(3):369–378

Krause PC, Wasynczuk O, Sudhoff SD, Pekarek SD (2013) Analysis of electric machinery and drive systems, vol 75. Wiley, New York

Kundur P, Balu NJ, Lauby MG (1994) Power system stability and control, vol 7. McGraw-Hill, New York, p 623

Schauder C, Mehta H (1993, July) Vector analysis and control of advanced static VAR compensators. In: IEE Proceedings C (generation, transmission and distribution), vol 140, no 4. IET Digital Library, pp 299–306

Chapter 4
Introduction to Model Predictive Control

Abstract and Brief Introduction to MPC Model predictive control (MPC) is one of the advanced techniques in industrial control that is gaining more focus in the modern optimal control literature. So far, it has found a lot of real-time applications in many important industries and was successfully implemented as an online controller (Fernandez-Camacho and Bordons-Alba, Model predictive control in the process industry, Springer, London, 1995). This chapter provides a glimpse of its functioning styles and designs to the readers.

Keywords Control horizon · Linear model-based MPC · Nonlinear model-based MPC · Prediction horizon · State-space model

4.1 MPC Description

A mathematical model of a system is an essential and core element for the MPC formulation. MPC uses the mathematical model of the system to predict the system behaviour within a time span starting from the present moment (present time instant). This time span containing the predicted behaviour is called the prediction horizon. The predicted behaviour within the prediction horizon is formulated in terms of the current system information (current state information) and future control trajectories within a time span starting from the present moment. This time span of the future control trajectories is called the control horizon. The length of the control horizon may or may not be equal to the prediction horizon. But it can never be more than that of the prediction horizon. Using this predicted behaviour, an objective function is formulated, which is optimized subjected to a set of system constraints and input constraints. The solving of the MPC optimal control problem generates the optimal control trajectories within the control horizon that optimize the future behaviour of the system within the prediction horizon. Then MPC applies the receding horizon principle (Wang 2009; Camacho and Alba 2013). The receding horizon principle states that in the entire optimal control trajectories within the control horizon, those control inputs corresponding to the present time instant are applied to the system, and the rest of the trajectories are neglected. When the next time instant arrives, the

P. Vidyasagar and K. Shanti Swarup, *Design and Development of Model Predictive Primary Control of Micro Grids*, Springer Tracts in Electrical and Electronics Engineering, https://doi.org/10.1007/978-981-19-5852-6_4

whole procedure of future prediction, optimal control problem formulation, optimal control trajectories generation, and application of receding horizon is repeated.

4.2 Advantages of MPC

It is worth discussing the superiority of the MPC compared to the other modern and conventional control techniques. The major advantages of the MPC are as follows (Fernandez-Camacho and Bordons-Alba 1995; Wang 2009; Camacho and Alba 2013):

- Optimal control: The MPC belongs to the class of optimal controllers like LQR. By this, we mean the MPC-generated control trajectories at each sample are a result of solving an optimization problem. Hence the generated control trajectories lead to minimization/maximization of some technical aspect or cost aspect of the system.
- Online controller: The MPC is employed as an online controller. It is not tuned offline for a range of operating conditions and scenarios that could exist in the system. It is dynamic in terms of tuning to different operating conditions, topological changes in the system, and changes in the constraint limits. Its ability to formulate the control problem based on the present state information and solve the problem online at every sampling instant puts it in a different league compared to the other conventional controllers, which are tuned offline.
- Constraint handling ability: Unlike other optimal controllers like LQR, the MPC is very good at handling the operational and security constraints of the system. It is this aspect that makes it special among the optimal controllers.
- Predictive nature: The most significant and unique aspect of MPC is its ability to predict future behaviour using the mathematical model and present state information. The control decisions are taken based on the optimized future behaviour. This allows the controller to look into future uncertainties while making the present decisions. Such a methodology obviously improves the system's reliability and stability. It allows the system to be better prepared for future surprises.
- Nonlinear applications: Although the MPC is initially meant for linear applications, it is successfully extended to many nonlinear applications and has been found to be very good in handling the system's nonlinearity.

4.3 MPC Types

An overview of different MPC types is shown in Fig. 4.1. A detailed explanation of each of these MPC types is given below:

- Based on the model structure:

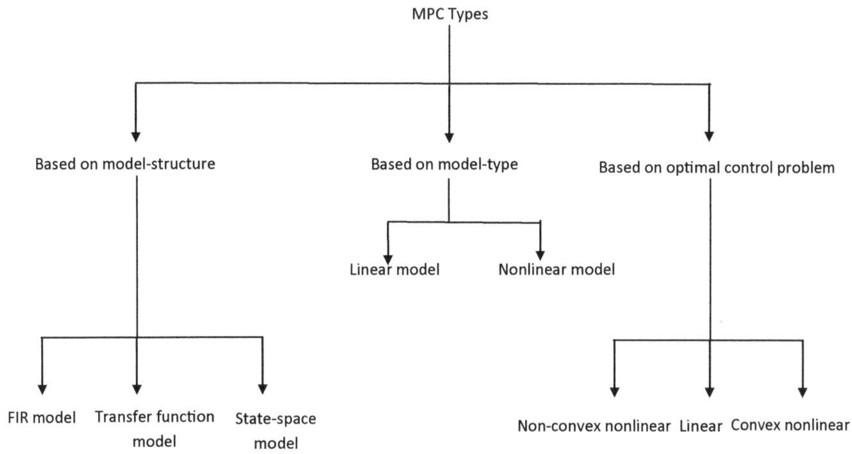

Fig. 4.1 MPC types

(a) Finite impulse response models: FIR models are the initial model structures used in the MPC formulation. The FIR models give in-depth information about the system gains, response time, and process delay. But a large number of response coefficients are required to construct the comprehensive behaviour of the system. Also, these structures are limited to stable systems. Dynamic Matrix Control (DMC) algorithms represent the FIR model-based MPC formulations.
(b) Transfer function models: The transfer function models are applicable to both stable and unstable systems. Generalized Predictive Control (GPC) algorithms represent the transfer function model-based MPC formulations. But the transfer function models cannot handle the multivariable systems effectively.
(c) State-space models: The modern MPC algorithms are based on the state-space models of the system. State-space models are applicable for both stable and unstable systems. They can handle multivariate systems. They can be applied to both SISO (single input and single output) systems and MIMO (multi-input and multi-output) systems. In this book, we are going to use state-space models for thc MPC formulations.

- Based on the model type:

(a) Linear model-based MPC: If the model of the system is linear, then the MPC using that model will be a linear model-based MPC. It is important to note that, although the system model is linear, the final optimal control problem of the MPC could be either a linear programming problem or a nonlinear optimization problem.
(b) Nonlinear model-based MPC: In this case, the model of the system is nonlinear, and the optimal control problem of the MPC is also nonlinear. However, the

optimal control problem may or may not be a convex one. This book deals with nonlinear model-based MPC formulations.

- Based on the complexity of the optimal control problem

(a) Linear control problem: The control problem of the MPC is a linear programming problem. This happens when the objective function and the constraints formulated in the MPC are linear. This kind of linear optimal control problem can be produced only by the system with a linear mathematical problem. It is important to note that the linear model of the system is not sufficient for creating the linear optimal control problem, but it is necessary.
(b) Convex nonlinear control problem: In this case, the optimal control problem is nonlinear. But it is a convex programming problem due to the fact that both objective function and constraints are all convex. Such kinds of problems can be solved in polynomial time complexity. Linear constrained quadratic programming problems (QP) and some of the quadratic constrained quadratic problems (QCQP) come under this category. These kinds of optimal control problems can be produced by both linear and nonlinear system models. This book deals with the convex optimal control problems of the MPC formulations.
(c) Non-convex nonlinear control problem: These kind of optimal control problems are NP-hard and are very difficult to solve online. A highly nonlinear mathematical model of a system produces this kind of non-convex nonlinear MPC control problems. The non-convexity can be due to objective function or constraints, or both. Later in the book, we can notice that the actual optimal control problem of the MPC is non-convex nonlinear in the case of micro-grids. This book mostly deals with how to solve such kinds of non-convex nonlinear optimal control problems in the micro-grids by approximating them to convex nonlinear control problems.

4.4 Linear Model-Based MPC/Linear-MPC/L-MPC

Now let us understand how to formulate a preliminary MPC for a system with the linear dynamic mathematical model. Such an MPC formulated from the linear mathematical model of the system is called "Linear model-based MPC" or "Linear-MPC" or simply L-MPC. As discussed in the previous section, the L-MPC can produce a linear optimal control problem or a convex nonlinear control problem. Out of the three model types discussed in the previous section, the state-space model of the system is preferred for the MPC formulation throughout the book. The formulation of the MPC involves three major steps, as shown in Fig. 4.2. The following basic steps will be followed throughout the book whenever an MPC is formulated:

(i) Formation of the augmented model
(ii) Formation of prediction vector
(iii) Formation of optimal control problem.

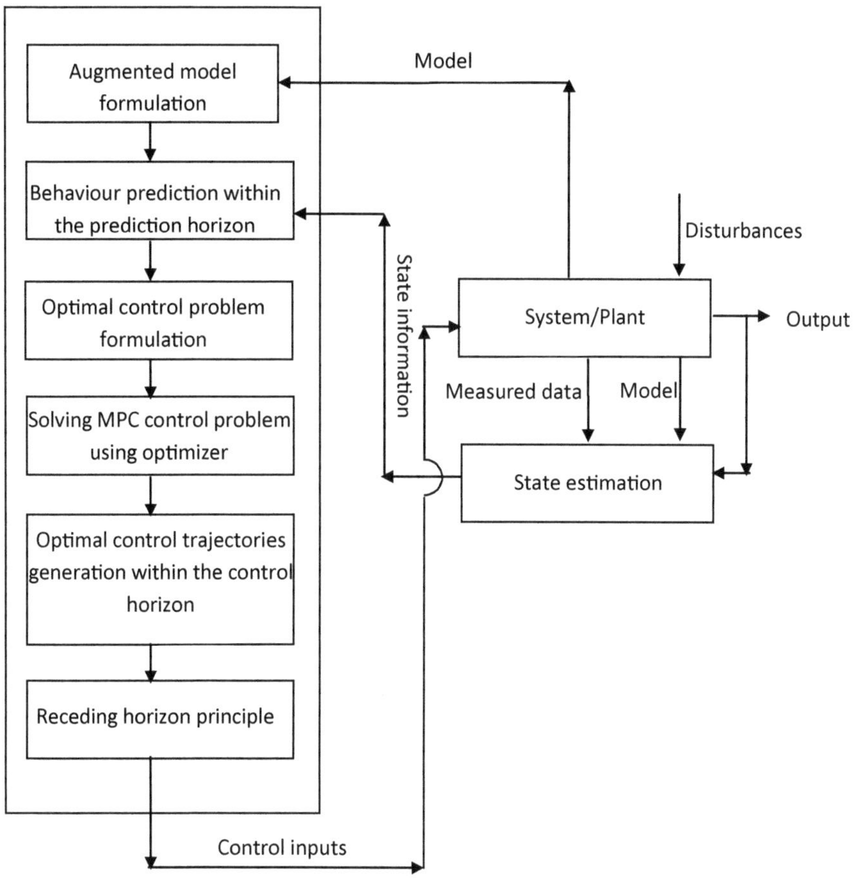

Fig. 4.2 Overview of the functioning of the MPC

4.4.1 Augmented Model

The complete linear mathematical model of a system is given by the following continuous-time state-space model:

$$\frac{\mathrm{d}\boldsymbol{Z}_c(t)}{\mathrm{d}t} = \boldsymbol{A}_c\boldsymbol{Z}_c(t) + \boldsymbol{B}_c\boldsymbol{u}(t)$$

$$\boldsymbol{Y}(t) = \boldsymbol{C}_c\boldsymbol{Z}_c(t) \tag{4.1}$$

$\boldsymbol{Z}_c$ is the continuous-time state vector with n states, $\boldsymbol{u}$ is the input vector with n_{ip} inputs, and $\boldsymbol{Y}$ is the output vector with n_{op} outputs. $(\boldsymbol{A}_c, \boldsymbol{B}_c, \boldsymbol{C}_c)$ is the continuous-time state-space triplet. At this point, it is important to understand two procedures regarding the design of MPC. The first procedure is designing the MPC in digital mode. In this mode, the MPC is formulated from the discrete model of the system. Hence the final optimal control problem is digital (or discrete, to be more precise). Then the discrete problem is solved using a discrete solver. The second procedure is designing the MPC in analog mode. In this mode, the MPC is formulated from the continuous-time model of the system. Hence the final optimal control problem is analog (or continuous, to be more precise). This continuous-time optimal problem is solved in the digital environment using a digital solver. It is important to note that in both cases, the final MPC problem solving is happening online in a digital environment. Hence the MPC is employed as a digital or discrete-time controller at the end. The only difference is whether it is designed in continuous time. Almost for all the linear system models, the MPC design is carried out in digital mode using the discrete-time model of the system. The continuous model (4.1) is converted to a discrete model using zero-order hold (ZOH):

$$\boldsymbol{Z}_c(k+1) = \boldsymbol{A}_d \boldsymbol{Z}_c(k) + \boldsymbol{B}_d \boldsymbol{u}(k)$$

$$\boldsymbol{Y}(k) = \boldsymbol{C}_d \boldsymbol{Z}_c(k) \tag{4.2}$$

$(\boldsymbol{A}_d, \boldsymbol{B}_d, \boldsymbol{C}_d)$ is the discrete-time state-space triplet. Most of the physical systems require an integral action in their control. The integral action deals with the control input increments rather than the control inputs. The integral action leads to the zero steady-state error. Hence, integrators are embedded into the system model (4.2) by converting the mathematical model of the system to an augmented model, as shown below:

$$\boldsymbol{Z}(k+1) = \boldsymbol{A}\boldsymbol{Z}(k) + \boldsymbol{B}\Delta\boldsymbol{u}(k)$$

$$\boldsymbol{Y}(k) = \boldsymbol{C}\boldsymbol{Z}(k) \tag{4.3}$$

$$\boldsymbol{Z}(k)^{\mathrm{T}} = \begin{bmatrix} \boldsymbol{Z}_c(k)^{\mathrm{T}} & \boldsymbol{u}(k-1)^{\mathrm{T}} \end{bmatrix},\ \boldsymbol{A} = \begin{bmatrix} \boldsymbol{A}_d & \boldsymbol{B}_d \\ 0_{n_{\mathrm{ip}}\times n} & \boldsymbol{I}_{n_{\mathrm{ip}}\times n_{\mathrm{ip}}} \end{bmatrix},\ \boldsymbol{B} = \begin{bmatrix} \boldsymbol{B}_d \\ \boldsymbol{I}_{n_{\mathrm{ip}}\times n_{\mathrm{ip}}} \end{bmatrix},$$
$$\boldsymbol{C} = \begin{bmatrix} \boldsymbol{C}_d & 0_{n_{\mathrm{op}}\times n_{\mathrm{ip}}} \end{bmatrix},\ \Delta\boldsymbol{u}(k) = \boldsymbol{u}(k) - \boldsymbol{u}(k-1)$$

4.4.2 Prediction Vector Within the Prediction Horizon

Let the time $t = t_s$ seconds be the present continuous-time moment at which the analysis is being carried out on the system. Let $k = k_i$ be the present sampling instant in the discrete-time domain. It is to be noted that the present sampling instant $k = k_i$ is given by $\{k = k_i = 0$, which implies that in continuous-time domain, we have $t = t_s\}$. Let $\boldsymbol{U}$ be a vector that contains the future control trajectories of the system from the present sampling instant $k = k_i$ to a control horizon of length N_c samples. Let $\boldsymbol{Y_E}$ be a vector with system outputs predicted from the sampling instant $k = k_i + 1$ to a prediction horizon of length N_p samples.

$$\boldsymbol{Y_E} = \left[\boldsymbol{Y}(k_i + 1)^{\mathrm{T}} \quad \boldsymbol{Y}(k_i + 2)^{\mathrm{T}} \ldots \boldsymbol{Y}(k_i + N_P)^{\mathrm{T}}\right]^{\mathrm{T}} \tag{4.4}$$

$$\boldsymbol{U} = \left[\Delta \boldsymbol{u}(k_i)^{\mathrm{T}} \quad \Delta \boldsymbol{u}(k_i + 1)^{\mathrm{T}} \ldots \Delta \boldsymbol{u}(k_i + N_C - 1)^{\mathrm{T}}\right]^{\mathrm{T}} \tag{4.5}$$

From (4.3), we can write the following equation:

$$\boldsymbol{Y}(k_i + m) = \boldsymbol{C}\boldsymbol{A}^m \boldsymbol{Z}(k_i) + \boldsymbol{C}\sum_{j=0}^{m-1} \boldsymbol{A}^{m-1-j}\boldsymbol{B}\Delta \boldsymbol{u}(k_i + j) \tag{4.6}$$

Substituting (4.6) in (4.4), we can get the following equation:

$$\boldsymbol{Y_E} = \boldsymbol{Y_{E1}}\boldsymbol{Z}(k_i) + \boldsymbol{Y_{E2}}\boldsymbol{U} \tag{4.7}$$

$$\boldsymbol{Y_{E1}} = \left[(\boldsymbol{C}\boldsymbol{A})^{\mathrm{T}} \quad (\boldsymbol{C}\boldsymbol{A}^2)^{\mathrm{T}} \ldots (\boldsymbol{C}\boldsymbol{A}^{N_p})^{\mathrm{T}}\right]^{\mathrm{T}}, \ \boldsymbol{Y_{E2}} = \left[\boldsymbol{F}_1^{\mathrm{T}} \quad \boldsymbol{F}_2^{\mathrm{T}} \quad \boldsymbol{F}_3^{\mathrm{T}} \ldots \boldsymbol{F}_{N_p}^{\mathrm{T}}\right]^{\mathrm{T}},$$

$$\boldsymbol{F}_j = \left[\boldsymbol{C}\boldsymbol{A}_{j,1}\boldsymbol{B} \quad \boldsymbol{C}\boldsymbol{A}_{j,2}\boldsymbol{B} \quad \boldsymbol{C}\boldsymbol{A}_{j,3}\boldsymbol{B} \ldots \boldsymbol{C}\boldsymbol{A}_{j,N_p}\boldsymbol{B}\right], \ \boldsymbol{A}_{j,l} = \left\{\begin{matrix} \boldsymbol{A}^{j-l} & \text{if} j \geq l \\ 0 & \text{if} j < l \end{matrix}\right\}$$

4.4.3 Optimal Control Problem Formulation

Once the predicted vector $\boldsymbol{Y_E}$ is formed from (4.7), an objective function is formulated using $\boldsymbol{Y_E}$ and is evaluated (minimized) subjected to a set of constraints on the system operation and security. Solving the optimal control problem generates the optimal control trajectories given by $\boldsymbol{U} = \boldsymbol{U_{opt}}$.

$$\begin{aligned} &\text{Min } J = (\boldsymbol{W} - \boldsymbol{Y_E})^{\mathrm{T}}(\boldsymbol{W} - \boldsymbol{Y_E}) + \boldsymbol{U}^{\mathrm{T}}\boldsymbol{R}\boldsymbol{U} \\ &\text{S. T.} \quad \boldsymbol{M}_1\boldsymbol{U} \leq \boldsymbol{N}_1, \\ &\boldsymbol{M}_2\boldsymbol{U} = \boldsymbol{N}_2 \end{aligned} \tag{4.8}$$

$\boldsymbol{W}$ is a vector with future set-points for the outputs. $\boldsymbol{R}$ is a positive definite weight matrix on the incremental control inputs. $\boldsymbol{M_1}$ and $\boldsymbol{M_2}$ are the constant matrices corresponding to linear inequality and linear equality system operational and security constraints. $\boldsymbol{N_1}$and $\boldsymbol{N_2}$ are the vectors containing bounds on these constraints. These constraints can be easily expressed in terms of the control vector $\boldsymbol{U}$ in the case of L-MPC. The optimal control trajectories within the control horizon are obtained by solving (4.8).

$$\boldsymbol{U}_{\mathbf{opt}} = \left[\Delta \boldsymbol{u}_{\mathbf{opt}}(k_i)^{\mathrm{T}} \quad \Delta \boldsymbol{u}_{\mathbf{opt}}(k_i+1)^{\mathrm{T}} \ldots \Delta \boldsymbol{u}_{\mathbf{opt}}(k_i+N_C-1)^{\mathrm{T}}\right]^{\mathrm{T}} \tag{4.9}$$

Once $\boldsymbol{U}_{opt}$ is found, then according to the receding horizon principle, input increments corresponding to the present sampling instant $\Delta \boldsymbol{u}_{\mathbf{opt}}(k_i)$ in (4.9) are applied to the micro-grid actuators. The rest of the trajectories in $\boldsymbol{U}_{\mathbf{opt}}$ are neglected. When the next sampling instant arrives, the entire process represented by (4.1)–(4.9) is repeated. It is to be noted that the above optimal control problem (4.9) is a quadratic problem with linear constraints. Hence it is a convex nonlinear control problem that can be solved in polynomial time complexity.

4.5 Nonlinear Model-Based MPC/Nonlinear-MPC/N-MPC

Most physical systems in nature are nonlinear. Thus most physical systems require nonlinear model-based MPC (N-MPC). The power engineering field is not different from this. The detailed model of a power system or a micro-grid or a single distributed generator is nonlinear, as discussed in the previous chapter. Apart from the model, the system's operational and security constraints can also be highly nonlinear. Hence most of the N-MPC control problems are non-convex nonlinear problems. Now let us formulate a preliminary N-MPC for a system with nonlinear dynamic mathematical given by:

$$\frac{\mathrm{d}\boldsymbol{X}(t)}{\mathrm{d}t} = \boldsymbol{f}(\boldsymbol{X}(t), \boldsymbol{V}(t))$$

$$\boldsymbol{Y}(t) = \boldsymbol{g}(\boldsymbol{X}(t)) \tag{4.10}$$

$\boldsymbol{X}$ is the state vector with n states, $\boldsymbol{Y}$ is the output vector with n_{op} outputs, and $\boldsymbol{V}$ is the input vector with n_{ip} inputs. At this stage, it is necessary to understand that, unlike L-MPC, most of the N-MPC formulations are formulated in the continuous-time analog mode. With the above continuous-time nonlinear model (4.10), the optimal control problem of the N-MPC is non-convex as given below:

$$\begin{aligned}
&\text{Min } J = \int_{t=t_s}^{t=t_s+T_p} (\boldsymbol{W}(t) - \boldsymbol{Y}(t))^{\mathrm{T}}(\boldsymbol{W}(t) - \boldsymbol{Y}(t))\mathrm{d}t + \int_{t=t_s}^{t=t_s+T_c} \boldsymbol{V}(t)^{\mathrm{T}}\boldsymbol{R}\boldsymbol{V}(t)\mathrm{d}t\\
&\text{S.T. } \frac{\mathrm{d}\boldsymbol{X}(t)}{\mathrm{d}t} = \boldsymbol{f}(\boldsymbol{X}(t), \boldsymbol{V}(t)),\\
&\boldsymbol{Y}(t) = \boldsymbol{g}(\boldsymbol{X}(t)),\\
&\boldsymbol{H}_1(\boldsymbol{X}(t), \boldsymbol{V}(t)) \le \boldsymbol{L}_1,\\
&\boldsymbol{H}_2(\boldsymbol{X}(t), \boldsymbol{V}(t)) = \boldsymbol{L}_2,
\end{aligned}$$

$$
\begin{aligned}
&\boldsymbol{V}_{\min} \leq \boldsymbol{V}(t) \leq \boldsymbol{V}_{\max},\\
&\boldsymbol{X}_{\min} \leq \boldsymbol{X}(t) \leq \boldsymbol{X}_{\max}
\end{aligned} \tag{4.11}
$$

$\boldsymbol{W}$ is the output reference set-point vector. $\boldsymbol{R}$ is the input penalty matrix. t_s is the present continuous-time moment at which the analysis is being carried out on the system. T_p and T_c are the prediction and control horizon lengths in the continuous-time domain. $\boldsymbol{H}_1$ and $\boldsymbol{H}_2$ are the vectors containing inequality and equality constraints which can be either linear or nonlinear. $\boldsymbol{L}_1$ and $\boldsymbol{L}_2$ are the vectors with the bounds for these constraints. $\boldsymbol{V}_{\min}$ and $\boldsymbol{V}_{\max}$ are the minimum and maximum limits on the inputs. $\boldsymbol{X}_{\min}$ and $\boldsymbol{X}_{\max}$ are the minimum and maximum limits on the states. Solving the above optimal control problem (4.11) online is extremely difficult as, most of the time, it will be an NP-hard problem due to its non-convexity.

4.6 Brief Review of MPC in Power Engineering

The initial applications of the MPC in power engineering are related to optimal switching of the power electronic converters, load frequency control in multi-area power networks, and voltage control in the distribution networks (Jin et al. 2017; Rodriguez et al. 2012; Cortés et al. 2008; Ersdal et al. 2015; Valverde and Cutsem 2013). The next major application of MPC is at the primary control level of a single inverter-DG unit or synchronous-DG unit in grid-connected mode. In the case of a synchronous DG, MPC in the primary control level provides much better performance compared to the conventional PI controllers (Wen et al. 2015). MPC was employed for the output control of an inverter-DG unit so that it eliminates the two-loop control theory with PI or PR controllers and provides an optimal control strategy for the grid-connected inverter-DG unit in a very simple way. Since the DGs are grid connected, the main objective of MPC is to control the power flow between the grid and DG unit in an optimal manner. Frequency and voltage control at PCC are not considered control objectives and are considered to be fixed throughout the analysis. Also, few works considered a stiff DC source and hence neglected the DC side dynamics of the inverter. This particular assumption simplifies the mathematical model of the DG unit and results in a linear mathematical model for the inverter DG (Cortés et al. 2009). Prediction of DG behaviour within the prediction horizon with linear models is easy, and the linear extrapolation suggested in these literatures is valid in that sense. However, when input source dynamics are considered, the grid-connected inverter-DG model becomes nonlinear (Delfino et al. 2012; Mahmud et al. 2013). To solve such nonlinear MPC, output error prediction is approximated to its relative degree by Taylor's series of approximation using Lie derivatives in such a way that input–output feedback linearization is possible (Lalili et al. 2011).

From a single DG connected to a grid, the focus of MPC application is shifted to micro-grids. The first application of MPC in the micro-grids is at the secondary control level. MPC is considered to be one of the finest control strategies for secondary control of micro-grid in which energy management of micro-grid is a priority (Bella et al. 2016). Objectives like unit commitment, restoration of frequency and voltages to their nominal values, optimal power flow, economic load dispatch, and reduction of carbon emissions are considered in the MPC formulation (Ahumada et al. 2015; Olivares et al. 2015; Minchala-Avila et al. 2016; Solanki et al. 2018; Parisio et al. 2014). It is implemented in a centralized manner at the secondary control level (Olivares et al. 2014; Prodan and Zio 2014; Tan et al. 2013). The latest research trend shifted to the application of MPC at the primary control level of the micro-grid.

Most of the research on MPC at the primary control is intended for inverter micro-grids and is decentralized in nature, with each DG unit having its own local MPC. In grid-connected mode, the MPC should perform the pre-specified active and reactive power regulation of its DG unit irrespective of the system frequency and its bus voltage (Errouissi et al. 2015; Hu et al. 2014a, b). In standalone mode, MPC should be able to provide a stable voltage at the local bus at a pre-specified constant frequency (Hu et al. 2014b; Tavakoli et al. 2016) or the reference voltage and frequency set to it from the V-Q and P-f droop characteristics (Babqi and Etemadi 2017; John and Ping Lam 2017). The inverter models used by the MPC for the prediction are linear in the inverter micro-grids. The linear models are due to neglecting DC side source dynamics and assuming constant frequency within the prediction horizon. The MPC control problem with linear models is a quadratic programming problem with polynomial time complexity. The dynamics of interaction between the DG and the rest of the micro-grid network are not available to the local MPC in decentralized control. In such a scenario, large prediction horizons lead to inaccurate predictions. Hence the prediction horizon in decentralized control is limited to one or two samples. Few works concentrated on centralized MPC for inverter micro-grids (Tan et al. 2012). Since the micro-grid model considered is linear, the MPC control problem is again a quadratic programming problem. However, large prediction horizons are possible in the centralized MPC design.

However, the present-day generic micro-grids consist of synchronous DGs, inverter DGs, linear loads, and nonlinear loads. The mathematical model of such generic micro-grids is highly nonlinear. For such generic micro-grids, the conventional MPC designs (both centralized and decentralized) intended for inverter micro-grids create issues. The conventional decentralized MPC with small prediction horizons leads to transient oscillatory behaviour and significant settling times. The drastic difference in the time constants of the synchronous DG and inverter DG combined with the coordination deficiency among their local controllers leads to the overall degrading performance of the decentralized MPC. Hence a centralized MPC controller with a large prediction horizon is still the best choice for such micro-grids (Liang et al. 2017). The centralized MPC is implemented either as a single controller or distributed with a group of buses coming under a centralized MPC (Yang et al. 2016; Pahasa and Ngamroo 2014). However, the mathematical model of such generic micro-grids is highly nonlinear. When such a model is used in the MPC for prediction, the control problem becomes a non-convex nonlinear programming problem that either fails to converge or is challenging to solve online with rapid sampling rates of the controller (Liu et al. 2019). This is the major drawback in the implementation of the centralized MPC at the primary control level of a micro-grid with highly nonlinear dynamics. Few literatures proposed linearized MPC using small-signal models of the generic micro-grids with highly nonlinear dynamics. The MPC solutions in these literatures are suboptimal at times during the transient period, but the online application of centralized MPC is ensured (Liu et al. 2019; Wang et al. 2014; Hossain et al. 2015). However, these studies are limited to small-signal analysis, and their capabilities during large-signal disturbances are a point of concern. Hence there is a need for an effective centralized MPC for small-signal and large-signal disturbances in the micro-grids at the primary control level. Simultaneously, the MPC should be computationally viable for online applications with large prediction and control horizons and should fit into the existing robustness studies for parametric uncertainties, state estimation errors, and unknown noise (Wang et al. 2014; Hossain et al. 2015).

4.7 Micro-grid MPC Methodologies Discussed in the Book

MPC in micro-grid primary control is mostly decentralized in nature with each DG unit having its own local MPC. But as discussed above decentralized MPC is limited to small prediction horizons mostly one or two samples. Only linear models of inverter DGs and SG-DGs are employed in the local MPC formulation as they are fine for prediction horizons of one or two samples. Within the prediction horizon, these linear models form a predicted behaviour that results in a convex quadratic optimization problem at each sampling instant which can be easily solved online since it exhibits a polynomial time complexity (Cimini and Bemporad 2017). In a generic micro-grid with both inertia- and non-inertia-based DGs and loads, the decentralized MPC leads to oscillatory transient behaviour and large settling times. Hence, there is a necessity for applying MPC as a single centralized primary controller for a micro-grid with both inertia- and non-inertia-based DGs and loads. The centralized MPC allows larger prediction horizons and provides excellent transient and steady-state properties compared to decentralized MPC.

However, when large prediction horizons are employed, the nonlinearity of the micro-grid model has to be considered. Otherwise the predicted micro-grid behaviour will become inaccurate. With such a nonlinear model of the micro-grid, the optimal control problem of the centralized MPC becomes a non-convex nonlinear programming optimization problem. The optimal control problem has to be solved online at rapid sampling rate requirements of the primary controller (Daoud and Fernando 2011). Such a non-convex nonlinear programming problem either cannot converge or extremely difficult to be solved in milliseconds range making the controller not viable for online implementation.

- **LTI-MPC**:
  Chapter 5 of the book discusses the linear time-invariant MPC (LTI-MPC) for the nonlinear micro-grid control. At each sample, the LTI-MPC linearizes the nonlinear micro-grid model around the operating point corresponding to that sample. This results in a linear time-invariant (LTI) model of the micro-grid. This LTI model is then used to predict the micro-grid response within the prediction horizon at that sample. The original non-convex nonlinear programming problem can be approximated to be a convex quadratic programming problem. LTI-MPC formulation is fairly simple and encourages rapid sampling rates. It is viable and easy to implement online.
- **LTV-MPC**:
  The LTI model is generally inaccurate and closely approximates the nonlinear model only in the small neighbourhood of the operating point. The LTI model deviates from the original nonlinear model with an increase in the length of the prediction horizon N_p. Hence LTI-MPC is limited to small prediction horizon lengths. When a load disturbance occurs in the micro-grid, LTI-MPC indeed controls the micro-grid and moves it towards the new steady-state. However, it leads to large oscillatory transient responses and large settling times due to the highly inaccurate prediction. Also, due to the receding horizon principle, control inputs that correspond to the present sample are chosen from the optimal control trajectories generated at each sample. The rest of the trajectories ("tail") are neglected. This is a waste of information that is optimal in some sense.

  Chapter 6 of the book discussed a linear time-variant MPC (LTV-MPC) for nonlinear micro-grid control that can tackle the issues in LTI-MPC. At each sample, within the prediction horizon, the nonlinear micro-grid model is linearized around the state and input reference trajectories. The linearization results in an LTV model of the micro-grid using which the forced response of the micro-grid is predicted. The natural response of the micro-grid within the prediction horizon is predicted by solving the nonlinear micro-grid model along the state and input reference trajectories. The sum of the forced and natural responses forms the complete predicted response of the micro-grid within the prediction horizon. Such a predicted response closely matches the actual nonlinear model and also allows higher prediction horizons. The actual non-convex nonlinear programming problem of the nonlinear MPC is approximated to a convex quadratic programming problem in the LTV-MPC. Due to a good prediction model, the LTV-MPC exhibits better transient properties and needs less settling times to reach the new steady-state whenever a disturbance occurs. Due to the receding horizon principle, inputs corresponding to the present time instant in the optimal control trajectories are applied to the micro-grid actuators. The rest of the optimal control trajectories are neglected. These neglected and unused trajectories form the "tail" of the LTV-MPC, which has optimal or suboptimal information. This "tail" computed in the present prediction horizon forms the input reference trajectories for the next prediction horizon when the following sample arrives. This avoids the waste of information calculated and further improves the settling time of the controller.

- **Special Functions**:
  The number of optimizing variables in the optimal control problem of the MPC (either LTI-MPC or LTV-MPC) depends on the number of inputs and the number of pulses to capture the control trajectory of each input. The number of pulses to capture the trajectory of each input is equal to the length of control horizon N_c. Thus the number of optimizing variables in MPC increases with an increase in the number of inputs and length of the control horizon. This is again an obstacle for applying it online as a centralized controller for micro-grids. To deal with this issue, Chapter 7 discusses the application of special orthonormal basis functions, namely Laguerre functions and two-parameter Kautz functions in the MPC formulation. The main theory behind this is to approximate the control trajectory of each input with a linear combination of the finite number of special orthonormal basis functions. The lesser the number of functions to approximate the control trajectory of each input to a reasonable accuracy more is the possibility of applying the proposed MPC online. The choice of type of orthonormal basis functions depends on the oscillatory behaviour of the control trajectory that is to be approximated. The Laguerre network consists of a real pole and has a very simple structure that can be easily generated. However, when the control trajectory is highly oscillatory, then a two-parameter Kautz network that has a pair of complex conjugate poles in its structure is very useful for the approximation.
- **Disturbance Compensator**:
  Chapter 8 focussed on incorporating the robust characteristics into the LTI-MPC and LTV-MPC by designing a disturbance compensator. This enables the MPC to deal with the parametric uncertainties, modelling errors, state estimation errors, and unknown disturbances.

4.8 Key Takeaways

- MPC uses the mathematical model of the system to predict the system behaviour and uses this predicted behaviour to formulate an objective function, which is optimized subjected to a set of system constraints and input constraints to generate the optimal control trajectories.
- Online optimal control, constraint handling capability, predictive nature, and suitability to the nonlinear systems are the major advantages it brings to the table.
- MPC can use either the FIR model, transfer function model, or state-space model of the system for the formulation of the optimal control problem.
- Based on the model type, the MPC can be either linear model-based or nonlinear model-based MPC.
- Based on the optimal control problem, the MPC can be either linear, convex nonlinear, or non-convex nonlinear programming-based MPC.

References

Ahumada C, Cárdenas R, Saez D, Guerrero JM (2015) Secondary control strategies for frequency restoration in islanded microgrids with consideration of communication delays. IEEE Trans Smart Grid 7(3):1430–1441

Babqi AJ, Etemadi AH (2017) MPC-based microgrid control with supplementary fault current limitation and smooth transition mechanisms. IET Generation Transmission Distribution 11(9):2164–2172

La Bella A, Cominesi SR, Sandroni C, Scattolini R (2016) Hierarchical predictive control of microgrids in islanded operation. IEEE Trans Autom Sci Eng 14(2):536–546

Camacho EF, Alba CB (2013) Model predictive control. Springer Science & Business Media

Cimini G, Bemporad A (2017) Exact complexity certification of active-set methods for quadratic programming. IEEE Trans Autom Control 62(12):6094–6109

Cortés P, Kazmierkowski MP, Kennel RM, Quevedo DE, Rodríguez J (2008) Predictive control in power electronics and drives. IEEE Trans Industr Electron 55(12):4312–4324

Cortés P, Ortiz G, Yuz JI, Rodríguez J, Vazquez S, Franquelo LG (2009) Model predictive control of an inverter with output LC filter for UPS applications. IEEE Trans Industr Electron 56(6):1875–1883

Daoud M, Fernando X (2011) On the communication requirements for the smart grid. Energy Power Eng 3(01):53

Delfino F, Denegri GB, Invernizzi M, Procopio R (2012) Feedback linearisation oriented approach to Q-V control of grid connected photovoltaic units. IET Renew Power Gener 6(5):324–339

Errouissi R, Muyeen SM, Al-Durra A, Leng S (2015) Experimental validation of a robust continuous nonlinear model predictive control based grid-interlinked photovoltaic inverter. IEEE Trans Industr Electron 63(7):4495–4505

Ersdal AM, Imsland L, Uhlen K (2015) Model predictive load-frequency control. IEEE Trans Power Syst 31(1):777–785

Fernandez-Camacho E, Bordons-Alba C (1995) Model predictive control in the process industry. Springer, London

Hossain MJ, Mahmud MA, Milano F, Bacha S, Hably A (2015) Design of robust distributed control for interconnected microgrids. IEEE Trans Smart Grid 7(6):2724–2735

Hu J, Zhu J, Dorrell DG (2014a) Model predictive control of grid-connected inverters for PV systems with flexible power regulation and switching frequency reduction. IEEE Trans Ind Appl 51(1):587–594

Hu J, Zhu J, Dorrell DG (2014b) Model predictive control of inverters for both islanded and grid-connected operations in renewable power generations. IET Renew Power Generation 8(3):240–248

John T, Ping Lam S (2017) Voltage and frequency control during microgrid islanding in a multi-area multi-microgrid system. IET Generation Transmission Distribution 11(6):1502–1512

Jin N, Hu S, Gan C, Ling Z (2017) Finite states model predictive control for fault-tolerant operation of a three-phase bidirectional AC/DC converter under unbalanced grid voltages. IEEE Trans Industr Electron 65(1):819–829

Lalili D, Mellit A, Lourci N, Medjahed B, Berkouk EM (2011) Input output feedback linearization control and variable step size MPPT algorithm of a grid-connected photovoltaic inverter. Renew Energy 36(12):3282–3291

Liang L, Hou Y, Hill DJ (20170 Design guidelines for MPC-based frequency regulation for islanded microgrids with storage, voltage, and ramping constraints. IET Renew Power Generation 11(8):1200–1210

Liu K, Liu T, Tang Z, Hill DJ (2019) Distributed MPC-based frequency control in networked microgrids with voltage constraints. IEEE Trans Smart Grid 10(6):6343–6354

Mahmud MA, Pota HR, Hossain MJ, Roy NK (2013) Robust partial feedback linearizing stabilization scheme for three-phase grid-connected photovoltaic systems. IEEE J Photovoltaics 4(1):423–431

Minchala-Avila LI, Garza-Castanon L, Zhang Y, Ferrer HJA (2016) Optimal energy management for stable operation of an islanded microgrid. IEEE Trans Industr Inf 12(4):1361–1370
Olivares DE, Cañizares CA, Kazerani M (2014) A centralized energy management system for isolated microgrids. IEEE Trans Smart Grid 5(4):1864–1875
Olivares DE, Lara JD, Cañizares CA, Kazerani M (2015) Stochastic-predictive energy management system for isolated microgrids. IEEE Trans Smart Grid 6(6):2681–2693
Pahasa J, Ngamroo I (2014) Coordinated control of wind turbine blade pitch angle and PHEVs using MPCs for load frequency control of microgrid. IEEE Syst J 10(1):97–105
Parisio A, Rikos E, Glielmo L (2014) A model predictive control approach to microgrid operation optimization. IEEE Trans Control Syst Technol 22(5):1813–1827
Prodan I, Zio E (2014) A model predictive control framework for reliable microgrid energy management. Int J Electr Power Energy Syst 61:399–409
Rodriguez J, Kazmierkowski MP, Espinoza JR, Zanchetta P, Abu-Rub H, Young HA, Rojas CA (2012) State of the art of finite control set model predictive control in power electronics. IEEE Trans Industr Inf 9(2):1003–1016
Solanki BV, Cañizares CA, Bhattacharya K (2018) Practical energy management systems for isolated microgrids. IEEE Trans Smart Grid 10(5):4762–4775
Tan KT, Peng XY, So PL, Chu YC, Chen MZ (2012) Centralized control for parallel operation of distributed generation inverters in microgrids. IEEE Trans Smart Grid 3(4):1977–1987
Tan KT, So PL, Chu YC, Chen MZQ (2013) Coordinated control and energy management of distributed generation inverters in a microgrid. IEEE Trans Power Delivery 28(2):704–713
Tavakoli A, Negnevitsky M, Muttaqi KM (2016) A decentralized model predictive control for operation of multiple distributed generators in an islanded mode. IEEE Trans Ind Appl 53(2):1466–1475
Valverde G, Van Cutsem T (2013) Model predictive control of voltages in active distribution networks. IEEE Trans Smart Grid 4(4):2152–2161
Wang Y, Chen Z, Wang X, Tian Y, Tan Y, Yang C (2014) An estimator-based distributed voltage-predictive control strategy for AC islanded microgrids. IEEE Trans Power Electronics 30(7):3934–3951
Wang L (2009) Model predictive control system design and implementation using MATLAB®. Springer Science & Business Media
Wen G, Hu G, Hu J, Shi X, Chen G (2015) Frequency regulation of source-grid-load systems: a compound control strategy. IEEE Trans Industr Inf 12(1):69–78
Yang F, Sun Q, Han QL, Wu Z (2016) Cooperative model predictive control for distributed photovoltaic power generation systems. IEEE J Emerging Selected Topics Power Electronics 4(2):414–420

Chapter 5
LTI-MPC for the Micro-grid Control

Abstract The micro-grid models in general are highly nonlinear. A centralized model predictive controller (MPC) for the primary control of micro-grids is a wonderful choice in terms of performance. However, the optimal control problem of a centralized MPC with a nonlinear model of the micro-grid is a non-convex nonlinear optimization problem. Solving such a problem online is very difficult with the rapid sampling rate requirements of the primary control level of a micro-grid. To deal with the non-convex nature of the optimal control problem, an MPC design that uses a linear time-invariant (LTI) approximation of the nonlinear micro-grid model is discussed in this chapter. The MPC design is referred to as linear time-invariant MPC (LTI-MPC). The optimal control problem of this LTI-MPC is a quadratic programming problem, which is convex with polynomial time complexity. The inspiration for the LTI-MPC is drawn from the references (Megias et al. in 1999 European control conference (ECC), pp 2707–2712, 1999; Blet et al. in IFAC Proc 35(1)L147–152, 2002; Ławryńczuk in Int J Appl Math Comput Sci 25(4):833–847, 2015; Vidyasagar and Swarup in 2016 National Power Systems Conference (NPSC), pp 1–6, 2016; Sagar and Swarup in 2016 IEEE International Conference on Power Electronics, Drives and Energy Systems (PEDES), pp 1–6, 2016) given at the end of the chapter.

Keywords Frequency control · Linear time-invariant approximation · Load disturbance · One at a time sensitivity · Source intermittency · Voltage control

5.1 Mathematical Formulation of the LTI-MPC

At each sampling instant, the LTI-MPC uses an LTI approximation of the nonlinear model of the micro-grid within the corresponding prediction horizon. The LTI model is formed by linearizing the nonlinear model around the operating point that corresponds to that particular sampling instant. The operating point may or may not be stationary. This LTI model is used for the prediction of the micro-grid behaviour within the respective prediction horizon. The predicted behaviour is then used in the formulation of the optimal control problem of the LTI-MPC, which is a convex quadratic programming problem, unlike the original non-convex nonlinear

P. Vidyasagar and K. Shanti Swarup, *Design and Development of Model Predictive Primary Control of Micro Grids*, Springer Tracts in Electrical and Electronics Engineering, https://doi.org/10.1007/978-981-19-5852-6_5

programming problem in the case of a nonlinear MPC (Vidyasagar and Swarup 2018).

5.1.1 Augmented Model

The complete nonlinear model of the micro-grid is given by the following continuous-time state-space model:

$$\begin{aligned}\frac{\mathrm{d}\boldsymbol{X}(t)}{\mathrm{d}t} &= \boldsymbol{f}(\boldsymbol{X}(t), \boldsymbol{V}(t))\\ \boldsymbol{Y}(t) &= \boldsymbol{g}(\boldsymbol{X}(t))\end{aligned} \tag{5.1}$$

$\boldsymbol{X}$ is the continuous-time state vector with n states, $\boldsymbol{V}$ is the input vector with n_{ip} inputs, and $\boldsymbol{Y}$ is the output vector with n_{op} outputs. Let the time $t = t_s$ seconds be the present time moment at which the analysis is being carried out. The linear time-invariant approximation of the continuous-time nonlinear model (5.1) is given by the Taylor's series expansion of (5.1) around the operating point $(\boldsymbol{X}(t_s), \boldsymbol{V}(t_s - T_s))$ with the higher-order terms neglected. The operating point can be either stationary or non-stationary. T_s is the sampling time of the controller. The LTI approximation is given by:

$$\begin{aligned}\frac{\mathrm{d}(\Delta\boldsymbol{X}(t))}{\mathrm{d}t} &= \boldsymbol{f}(\boldsymbol{X}(t_s), \boldsymbol{V}(t_s - T_s))\\ &\quad + \frac{\partial \boldsymbol{f}}{\partial \boldsymbol{X}}\bigg|_{\substack{X = X(t_s)\\ V = V(t_s - T_s)}} \Delta\boldsymbol{X}(t) + \frac{\partial \boldsymbol{f}}{\partial \boldsymbol{V}}\bigg|_{\substack{X = X(t_s)\\ V = V(t_s - T_s)}} \Delta\boldsymbol{V}(t)\\ \boldsymbol{Y}(t) &= \boldsymbol{g}(\boldsymbol{X}(t_s)) + \frac{\partial \boldsymbol{g}}{\partial \boldsymbol{X}}\bigg|_{\substack{X = X(t_s)\\ V = V(t_s - T_s)}} \Delta\boldsymbol{X}(t)\\ \Delta\boldsymbol{V}(t) &= \boldsymbol{V}(t) - \boldsymbol{V}(t_s - T_s),\ \Delta\boldsymbol{X}(t) = \boldsymbol{X}(t) - \boldsymbol{X}(t_s)\end{aligned} \tag{5.2}$$

The linear model (5.2) can be written as:

$$\begin{aligned}\frac{\mathrm{d}\boldsymbol{Z}_c(t)}{\mathrm{d}t} &= \overline{\boldsymbol{Z}} + \boldsymbol{A}_t\boldsymbol{Z}_c(t) + \boldsymbol{B}_t\boldsymbol{u}(t)\\ \boldsymbol{Y}(t) &= \overline{\boldsymbol{Y}} + \boldsymbol{C}_t\boldsymbol{Z}_c(t)\end{aligned} \tag{5.3}$$

$$\boldsymbol{Z}_c(t) = \Delta\boldsymbol{X}(t), \overline{\boldsymbol{Z}} = \boldsymbol{f}(\boldsymbol{X}(t_s), \boldsymbol{V}(t_s - T_s)), \overline{\boldsymbol{Y}} = \boldsymbol{g}(\boldsymbol{X}(t_s)), \boldsymbol{u}(t) = \Delta\boldsymbol{V}(t),$$

$$\boldsymbol{A}_t = \frac{\partial \boldsymbol{f}}{\partial \boldsymbol{X}}\bigg|_{\substack{X = X(t_s)\\ V = V(t_s - T_s)}}, \boldsymbol{B}_t = \frac{\partial \boldsymbol{f}}{\partial \boldsymbol{V}}\bigg|_{\substack{X = X(t_s)\\ V = V(t_s - T_s)}}, \boldsymbol{C}_t = \frac{\partial \boldsymbol{g}}{\partial \boldsymbol{X}}\bigg|_{\substack{X = X(t_s)\\ V = V(t_s - T_s)}}$$

The continuous model (5.3) is then converted to a discrete model using zero-order hold (ZOH):

$$\begin{aligned} \boldsymbol{Z}_c(k+1) &= \widetilde{\boldsymbol{Z}} + \boldsymbol{A}_d \boldsymbol{Z}_c(k) + \boldsymbol{B}_d \boldsymbol{u}(k) \\ \boldsymbol{Y}(k) &= \widetilde{\boldsymbol{Y}} + \boldsymbol{C}_d \boldsymbol{Z}_c(k) \end{aligned} \tag{5.4}$$

$(\boldsymbol{A}_d, \boldsymbol{B}_d, \boldsymbol{C}_d)$ is the discrete-time state-space triplet with $\{k = 0 \Leftrightarrow t = t_s\}$. $\widetilde{\boldsymbol{Z}}$, $\widetilde{\boldsymbol{Y}}$ are discrete counterparts to $\overline{\boldsymbol{Z}}$, $\overline{\boldsymbol{Y}}$. As the micro-grid control requires an integral action, the integrators are embedded into (5.4), converting it to an augmented model given by:

$$\begin{aligned} \boldsymbol{Z}(k+1) &= \overline{\boldsymbol{\theta}} + \boldsymbol{A}\boldsymbol{Z}(k) + \boldsymbol{B}\Delta\boldsymbol{u}(k) \\ \boldsymbol{Y}(k) &= \widetilde{\boldsymbol{Y}} + \boldsymbol{C}\boldsymbol{Z}(k) \end{aligned} \tag{5.5}$$

$$\overline{\boldsymbol{\theta}}^T = \begin{bmatrix} \widetilde{\boldsymbol{Z}}^T & 0_{1\times n_{ip}} \end{bmatrix}, \boldsymbol{Z}(k)^T = \begin{bmatrix} \boldsymbol{Z}_c(k)^T & \boldsymbol{u}(k-1)^T \end{bmatrix},$$
$$\boldsymbol{A} = \begin{bmatrix} \boldsymbol{A}_d & \boldsymbol{B}_d \\ 0_{n_{ip}\times n} & \boldsymbol{I}_{n_{ip}\times n_{ip}} \end{bmatrix}, \boldsymbol{B} = \begin{bmatrix} \boldsymbol{B}_d \\ \boldsymbol{I}_{n_{ip}\times n_{ip}} \end{bmatrix}$$
$$\boldsymbol{C} = \begin{bmatrix} \boldsymbol{C}_d & 0_{n_{op}\times n_{ip}} \end{bmatrix}, \quad \Delta\boldsymbol{u}(k) = \boldsymbol{u}(k) - \boldsymbol{u}(k-1)$$

5.1.2 Prediction Vector Within the Prediction Horizon

Let $\boldsymbol{U}$ be a vector that contains the future control trajectories from the present sampling instant $k = k_i$ to a control horizon of length N_c samples. Let $\boldsymbol{Y}_E$ be a vector with micro-grid outputs predicted from the sampling instant $k = k_i + 1$ to a prediction horizon of length N_p samples. It is to be noted that the present sampling instant $k = k_i$ is given by $\{k = k_i = 0 \Leftrightarrow t = t_s\}$.

$$\boldsymbol{Y}_E = \begin{bmatrix} \boldsymbol{Y}(k_i+1)^T & \boldsymbol{Y}(k_i+2)^T & \ldots \boldsymbol{Y}(k_i+N_P)^T \end{bmatrix}^T \tag{5.6}$$

$$\boldsymbol{U} = \begin{bmatrix} \Delta\boldsymbol{u}(k_i)^T & \Delta\boldsymbol{u}(k_i+1)^T \ldots \Delta\boldsymbol{u}(k_i+N_C-1)^T \end{bmatrix}^T \tag{5.7}$$

From (5.5) we can write the following equation:

$$\boldsymbol{Y}(k_i+m) = \widetilde{\boldsymbol{Y}} + \boldsymbol{C}\boldsymbol{A}^m\boldsymbol{Z}(k_i) + \boldsymbol{C}\sum_{j=0}^{m-1}\boldsymbol{A}^j\overline{\boldsymbol{\theta}} + \boldsymbol{C}\sum_{j=0}^{m-1}\boldsymbol{A}^{m-1-j}\boldsymbol{B}\Delta\boldsymbol{u}(k_i+j) \tag{5.8}$$

Substituting (5.8) in (5.6) we can get the following equation:

$$
\begin{aligned}
\boldsymbol{Y}_E &= \boldsymbol{Y}_1 + \boldsymbol{Y}_2\bar{\boldsymbol{\theta}} + \boldsymbol{Y}_3\boldsymbol{Z}(k_i) + \boldsymbol{Y}_4\boldsymbol{U} \\
\boldsymbol{Y}_1 &= \left[\widetilde{Y}^T \;\; \widetilde{Y}^T \;\; \widetilde{Y}^T \ldots \widetilde{Y}^T\right]^T, \\
\boldsymbol{Y}_2 &= \left[\boldsymbol{C}^T \; (\boldsymbol{C}+\boldsymbol{C}\boldsymbol{A})^T \ldots \left(\boldsymbol{C}+\boldsymbol{C}\boldsymbol{A}+\cdots+\boldsymbol{C}\boldsymbol{A}^{N_p-1}\right)^T\right]^T, \\
\boldsymbol{Y}_3 &= \left[(\boldsymbol{C}\boldsymbol{A})^T \; \left(\boldsymbol{C}\boldsymbol{A}^2\right)^T \ldots \left(\boldsymbol{C}\boldsymbol{A}^{N_p}\right)^T\right]^T, \\
\boldsymbol{Y}_4 &= \left[\boldsymbol{F}_1^T \; \boldsymbol{F}_2^T \; \boldsymbol{F}_3^T \ldots \boldsymbol{F}_{N_p}^T\right]^T \\
\boldsymbol{F}_j &= \left[\boldsymbol{C}\boldsymbol{A}_{j,1}\boldsymbol{B} \;\; \boldsymbol{C}\boldsymbol{A}_{j,2}\boldsymbol{B} \;\; \boldsymbol{C}\boldsymbol{A}_{j,3}\boldsymbol{B} \ldots \boldsymbol{C}\boldsymbol{A}_{j,N_p}\boldsymbol{B}\right], \\
\boldsymbol{A}_{j,l} &= \begin{Bmatrix} \boldsymbol{A}^{j-l} & \text{if } j \geq l \\ 0 & \text{if } j < l \end{Bmatrix}
\end{aligned} \tag{5.9}
$$

5.1.3 Optimal Control Problem Formulation

Once $\boldsymbol{Y}_E$ is formed from (5.9), an objective function is formulated using $\boldsymbol{Y}_E$ and is evaluated (minimized) subjected to a set of constraints on the micro-grid operation to generate the optimal control trajectories given by $\boldsymbol{U} = \boldsymbol{U}_{\mathbf{opt}}$.

$$
\begin{aligned}
&\text{Min } J = (\boldsymbol{W} - \boldsymbol{Y}_E)^T(\boldsymbol{W} - \boldsymbol{Y}_E) + \boldsymbol{U}^T\boldsymbol{R}\boldsymbol{U} \\
&\text{S.T.}\Delta\boldsymbol{u}_{\mathbf{min}} \leq \Delta\boldsymbol{u}(k_i) \leq \Delta\boldsymbol{u}_{\mathbf{max}} \\
&\quad \boldsymbol{u}_{\mathbf{min}} \leq \boldsymbol{u}(k_i) \leq \boldsymbol{u}_{\mathbf{min}} \\
&\quad \boldsymbol{V}_{\mathbf{min}} \leq \boldsymbol{V}(k_i) \leq \boldsymbol{V}_{\mathbf{min}} \\
&\quad \boldsymbol{M}_{\mathbf{eq}}\boldsymbol{U} = \boldsymbol{\alpha}_{\mathbf{eq}} \\
&\quad \boldsymbol{M}_{\mathbf{ineq}}\boldsymbol{U} \leq \boldsymbol{\alpha}_{\mathbf{ineq}}
\end{aligned} \tag{5.10}
$$

$\boldsymbol{W}$ is a vector with future set-points for the outputs. $\boldsymbol{R}$ is a positive definite weight matrix on input increments. $\boldsymbol{M}_{\mathbf{eq}}$ and $\boldsymbol{M}_{\mathbf{ineq}}$ are the matrices related to the equality and inequality operational constraints of the micro-grid. $\boldsymbol{\alpha}_{\mathbf{eq}}$ and $\boldsymbol{\alpha}_{\mathbf{ineq}}$ are the vectors containing the limits on the equality and inequality operational constraints of the micro-grid. The optimal control trajectories within the control horizon are obtained by solving (5.10).

$$
\boldsymbol{U}_{\mathbf{opt}} = \left[\Delta\boldsymbol{u}_{\mathbf{opt}}(k_i)^T \;\; \Delta\boldsymbol{u}_{\mathbf{opt}}(k_i+1)^T \ldots \Delta\boldsymbol{u}_{\mathbf{opt}}(k_i+N_C-1)^T\right]^T \tag{5.11}
$$

Once $\boldsymbol{U}_{\mathbf{opt}}$ is found, then according to the receding horizon principle, input increments corresponding to the present sampling instant $\Delta\boldsymbol{u}_{\mathbf{opt}}(k_i)$ in (5.11) are applied to the micro-grid actuators. The rest of the trajectories in $\boldsymbol{U}_{\mathbf{opt}}$ are neglected. When the next sampling instant arrives, the entire process represented by (5.1)–(5.11) is repeated, as shown in Fig. 5.1.

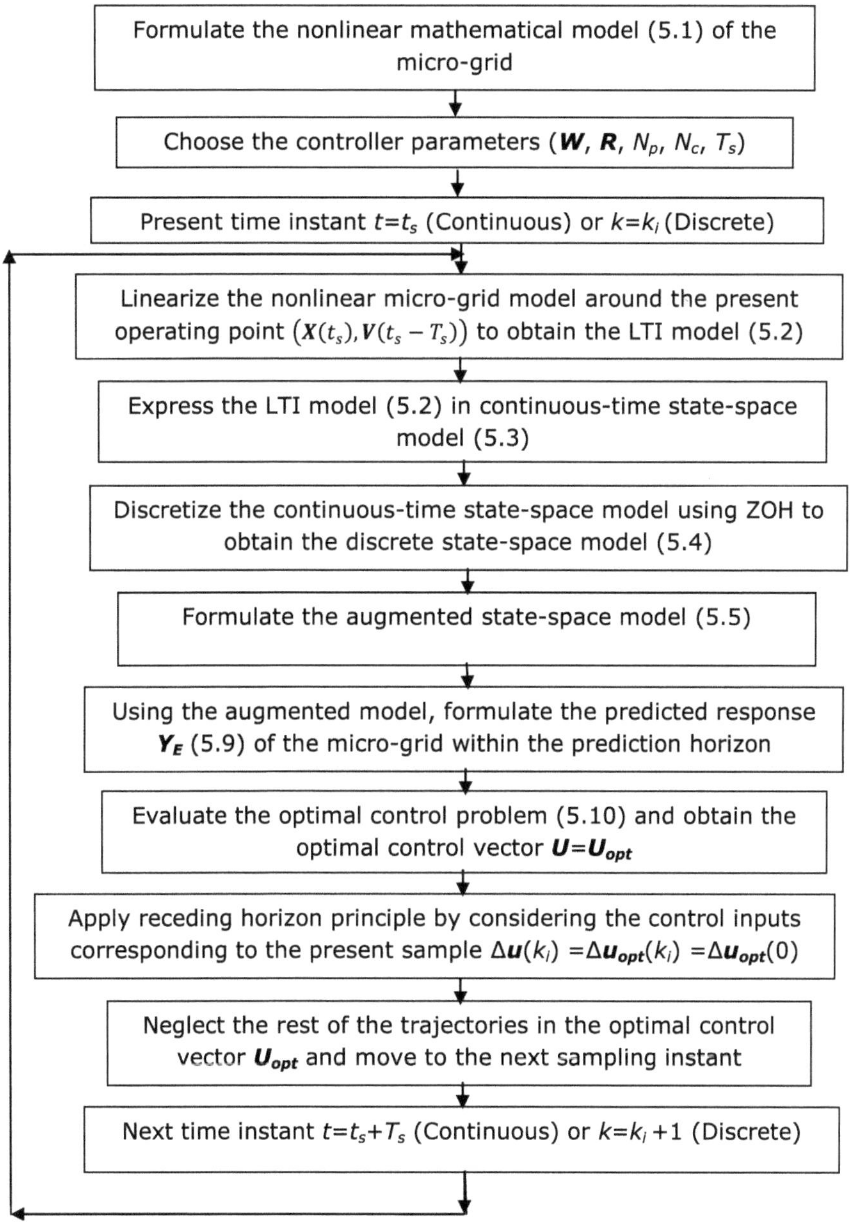

Fig. 5.1 Operation of the LTI-MPC

5.2 LTI-MPC for the Micro-grid Control

Now that the mathematical formulation part of the LTI-MPC was done, it is the time to test its performance capabilities in the micro-grid. In Chap. 3, we have a generic micro-grid shown in Fig. 3.1 was modelled in both grid-connected and standalone modes. Throughout the book, this micro-grid is used for the analysis of different MPC formulations. In particular, the standalone mode of the micro-grid was considered for the analysis as it is more challenging than the grid-connected mode. The analysis of MPC formulations in the grid-connected mode was left to the readers to investigate in the same lines as that of the standalone mode in the book. For the performance analysis, three kinds of load disturbances are created in the micro-grid from the initial operating point. They are (i) single R-L load disturbance, (ii) multiple R-L and induction motor (IM) load disturbances at a time, and (iii) EI-DG source intermittency (PV irradiance change). Detailed information about the initial operating point of the micro-grid can be found in the Appendix 1.

5.2.1 Role of Each DG Unit in the Micro-grid Control

During a disturbance, the SG-DG should control its bus voltage and system frequency during the transient period and restore them to their nominal values during the steady-state (Vidyasagar and Swarup 2018; Ranga et al. 2018). When it comes to the EI-DG, it should control its respective bus voltage during the transient period and restore it to its nominal value during the steady-state by optimally adjusting its reactive power injection (Vidyasagar and Swarup 2018). However, its participation in the system frequency control is rather arbitrary. If it does not participate in the frequency control, then its output active power injection is maintained constant at all times. The active power output, in this case, is the nominal power reference coming from the secondary control level, which generally is a maximum power point. If it participates in the frequency control, then a primary reserve margin should be provided to the EI-DG (Vidyasagar and Swarup 2018). The EI-DG uses this margin to participate in the frequency control during the transient period. However, in the steady-state, its active power output is restored to its nominal value. Hence in either of the two cases, only SG-DG meets the steady-state active power load changes and the corresponding losses in the network. In this book, the EI-DG is provided with a primary reserve margin of 15% of its maximum output power to make it participate in the transient frequency control. The choice of primary reserve margin is rather arbitrary in this book. However, in the real-time scenario, it is decided by the secondary control level after performing the necessary energy management optimization algorithms. Since we are using a photovoltaic (PV) source for the EI-DG, we refer the EI-DG in this book as PV-DG from now on for the convenience.

5.2.2 Operational Constraints

Apart from the input constraints given in the optimal control problem (5.10), the micro-grid shown in Fig. 3.1 has some operational constraints given below:

$$\begin{aligned} P_{G1,\min} &\leq P_{G1}(k_i+1) \leq P_{G1,\max} \\ Q_{G1,\min} &\leq Q_{G1}(k_i+1) \leq Q_{G1,\max} \\ P_{G3,\min} &\leq P_{G3}(k_i+1) \leq P_{G3,\max} \\ Q_{G3,\min} &\leq Q_{G3}(k_i+1) \leq Q_{G3,\max} \\ V_{\text{mppt}}\,(G,T) &\leq V_{\text{dc}}(k_i+1) \leq V_{\text{oc}}\,(G,T) \end{aligned}$$

(P_{G1}, Q_{G1}), (P_{G3}, Q_{G3}) are active and reactive powers of the SG-DG and PV-DG. The minimum and maximum limits of the active and reactive powers are given by $\left(P_{G1,\min}, P_{G1,\max}\right)$, $\left(Q_{G1,\min}, Q_{G1,\max}\right)$, $\left(P_{G3,\min}, P_{G3,\max}\right)$, $\left(Q_{G3,\min}, Q_{G3,\max}\right)$. $V_{\text{mppt}}(G,T)$ and $V_{\text{oc}}\,(G,T)$ are the output voltages of the PV array corresponding to the maximum power point and open circuit. V_{dc} is the DC-link capacitor voltage.

5.2.3 Choice of the Controller Parameters

- **Sampling time** (T_s): Lower the sampling time, the better will be the performance of the controller. However, the practicality of the controller implementation in real-time has to be taken into consideration. In this book, the sampling time is chosen to be $T_s = 1$ ms, as the millisecond's range is considered to be practical in the smart grid scenario.
- **Output set-point vector** $\boldsymbol{W}$: At each sample, within the prediction horizon, the set-point vector for the micro-grid outputs in the standalone mode as given in Chap. 3 in Sect. 3.7 is given by:

$$\boldsymbol{W}^T = \left[\boldsymbol{w}^T\,\boldsymbol{w}^T\,\boldsymbol{w}^T \ldots \boldsymbol{w}^T\right]_{1\times N_p n_{\text{op}}} \tag{5.12}$$

$$w^T = \left[377 \quad 1 \quad 1 \left(P_{3,\text{nom}} + \left(\frac{\omega_b - \omega_3}{\omega_b}\right)\frac{1}{R_{D3}}\right)\right] \tag{5.13}$$

- The reference set-point for the rotor speed ω_r is 377 rad/s (60 Hz). The reference set-point for the bus B1 voltage V_1 is 1 p.u. The reference set-point for the bus B3 voltage V_3 is 1 p.u. The PV-DG active power P_{G3} follows a droop characteristic. $P_{3,\text{nom}}$ is the nominal output of the PV-DG, which is set at 85% of the PV array's maximum output. The remaining 15% forms the primary reserve. R_{D3} is the droop constant of the PV-DG.
- **Prediction horizon length** (N_p): Since an LTI model is used for the output prediction within the prediction horizon, the upper cut off for N_p is decided based on

the value of N_p up to which the linearization is valid. For this reason, one at a time (OAT) sensitivity analysis is carried out for the worst-case load disturbance scenario. For the one at a time (OAT) sensitivity analysis, the input parameter is the length of the prediction horizon, and the output parameter is the error in the areas between the predicted response and actual response of the micro-grid within the prediction horizon. The output parameter is given by:

$$\text{Error} \quad (\varepsilon) = \int_{t_s}^{t_s+N_pT_s} \left|\left(y_{\text{actual}}(t) - y_{\text{predicted}}(t)\right)\right| \mathrm{d}t \tag{5.14}$$

where $t = t_s$ seconds be the present continuous-time instant whose discrete-time equivalent is given by $k = k_i$. The prediction horizon in the continuous-time domain is given by $t_s \leq t \leq t_s + N_pT_s$ and in the discrete-time domain is given by $k_i \leq k \leq k_i + N_p$. N_p is the length of the prediction horizon and T_s is the sampling time of the controller. $y_{\text{actual}}(t)$ is the actual response of the micro-grid within the prediction horizon. $y_{\text{predicted}}(t)$ is the predicted response of the micro-grid within the prediction horizon.

For the analysis, the extreme case disturbance of load hitting from minimum to sudden maximum is considered. The p.u minimum and maximum load parameters at different buses in the micro-grid shown in Fig. 3.1 of Chap. 3 are given in Appendix 1. At $t = 0.1$ s, the load at buses B5, B6, B7, and B8 is suddenly changed from their minimum values to maximum values. The error (ε) is found for different prediction horizon lengths. The error results are presented for different time instances in Fig. 5.2. From Fig. 5.2, we can observe that, for $1 \leq N_p \leq 164$, , the error (ε) is small at all times, and it keeps on decreasing as the time increases. In other words, the predicted response is converging more with the actual response as the time increases for $1 \leq N_p \leq 164$. But for $N_p \geq 165$, the error is increasing with time. This indicates that the MPC solution obtained using linearization of the micro-grid model is giving unstable results when it is applied for the actual nonlinear model. This indicates that the linearization is either not valid or highly inaccurate for $N_p \geq 165$. Hence the upper cut off for $N_p = 164$.

- The working range of N_p is $1 \leq N_p \leq 164$. From this working range, the required N_p can be chosen in such a way that it is as large as possible. However, while choosing N_p, we must consider the fact that the computational time of the control trajectories should be less than that of the sampling rate of the controller and there should be enough room for the communication delay. A prediction horizon of $N_p = 150$ is considered in this book as it satisfies the above requirements.
- **Control horizon length** (N_c): The control horizon length is less than or equal to the prediction horizon $N_c \leq N_p$. The advantage with a small control horizon is that it leads to a small number of decision variables in the optimal control problem at each sample. However, a large control horizon provides stable solutions and yields better results, as shown in Fig. 5.3. Figure 5.3 shows the micro-grid outputs for different control horizon lengths and a prediction horizon of length $N_p = 150$

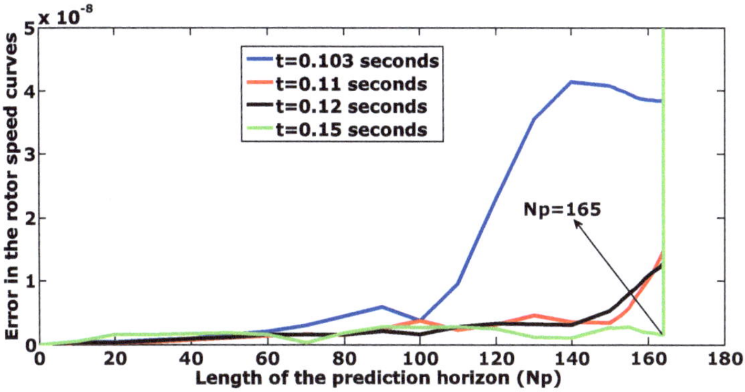

(a) Error in the rotor speed curves during the worst-case load disturbance

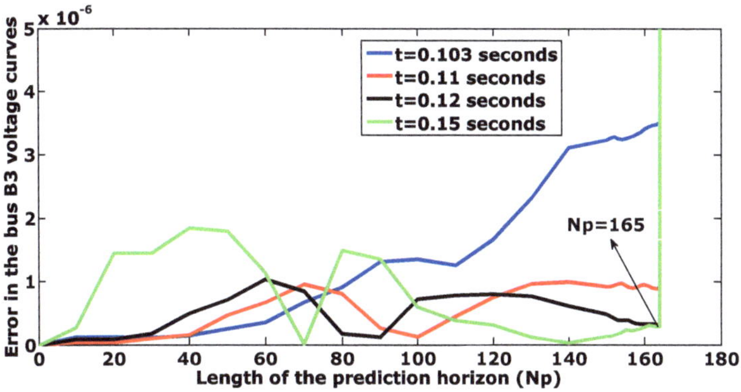

(b) Error in the bus B3 voltage curves during the worst-case load disturbance

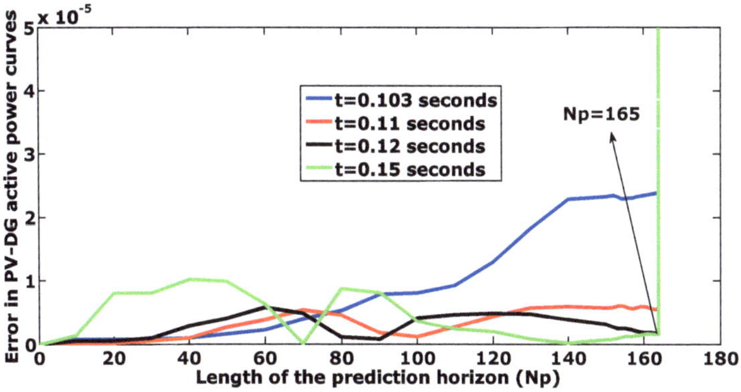

(c) Error in the PV-DG active power curves during the worst-case load disturbance

Fig. 5.2 Error for different prediction horizon lengths at different time instances during the worst-case load disturbance scenario: **a** rotor speed, **b** bus B3 voltage, **c** PV-DG active power, and **d** bus B1 voltage

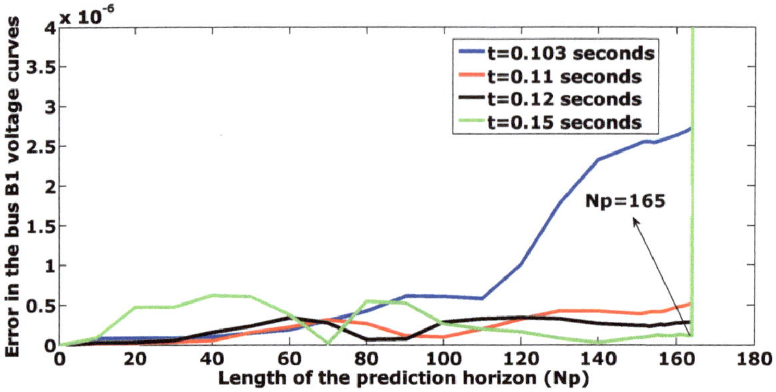

(d) Error in the bus B1 voltage curves during the worst-case load disturbance

Fig. 5.2 (continued)

during the worst-case load disturbance explained above. From the Fig. 5.3, we can observe that $N_p = N_c = 150$ gives the best performance as the settling time and peak deviations of the micro-grid outputs are less for $N_c = 150$ compared to the other values of N_c.

- **Penalty matrix on input increments ($\boldsymbol{R}$)**: The penalties on input increments should be as small as possible. Ideally, they should be zeros. However, zero penalties and very small penalties on input increments lead to numerical ill-conditioning of the optimal control problem (5.10). One of the significant indicators for the numerically ill-conditioned optimal control problem (5.10) is the very high condition number of the matrix $\boldsymbol{E} = \boldsymbol{Y}_4^T \boldsymbol{Y}_4 + \boldsymbol{R}$. During the worst-case load disturbance scenario, the maximum possible N_p and N_c found from the OAT analysis is given by $N_p = N_c = 164$. For this $N_p = N_c = 164$, the upper bound on the condition number of $\boldsymbol{E}$ is set to 1.5×10^{11} up to which the accuracy of the numerical calculations is satisfactory, and stable numerical solutions are produced. The upper bound is valid for any other N_p within the feasible range $1 \leq N_p \leq 164$. Based on this upper bound, the penalty on each SG-DG input is taken to be 0.0000001, and on each PV-DG input is considered to be 0.0001.

5.3 Performance Analysis

Example 5.1: Single R-L load disturbance

This is a case to represent the small-signal load disturbance in the micro-grid. At $t = 0.1$ s, the p.u-aggregated R-L load L7 at bus B7 ($R_{l7} + jX_{l7}$) of the micro-grid shown in Fig. 3.1 is suddenly changed from an initial value of 3.8306 + j3.8757 to 2.7948 + j2.8276 in the standalone mode. This change indicates a sudden load increase at bus B7. The performance of the LTI-MPC can be analysed using Fig. 5.4. To meet

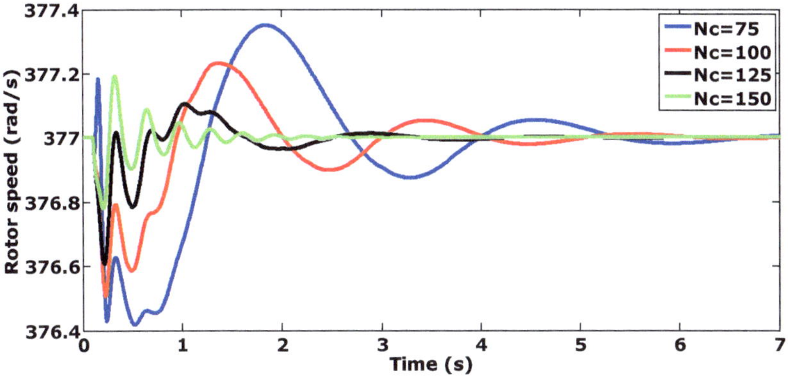

(a) Rotor speed during the worst-case load disturbance

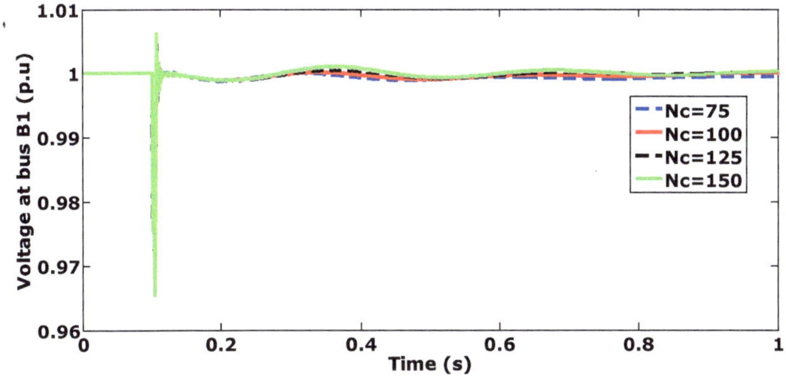

(b) Bus B1 voltage during the worst-case load disturbance

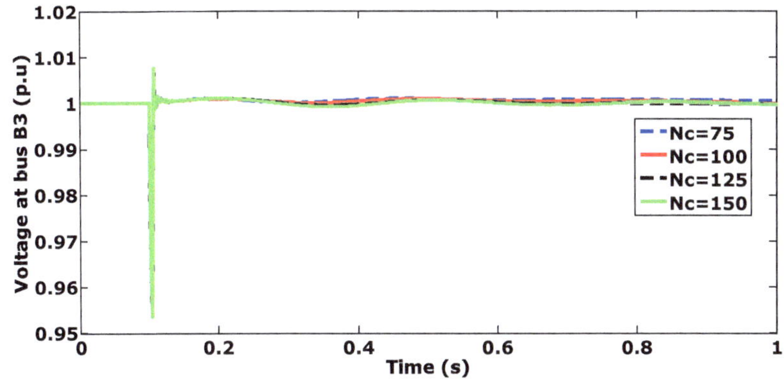

(c) Bus B3 voltage during the worst-case load disturbance

Fig. 5.3 LTI-MPC performance for different control horizon lengths and a prediction horizon length of $N_p = 150$, during the worst-case load disturbance scenario: **a** rotor speed, **b** bus B1 voltage, **d** bus B3 voltage, and **d** PV-DG active power

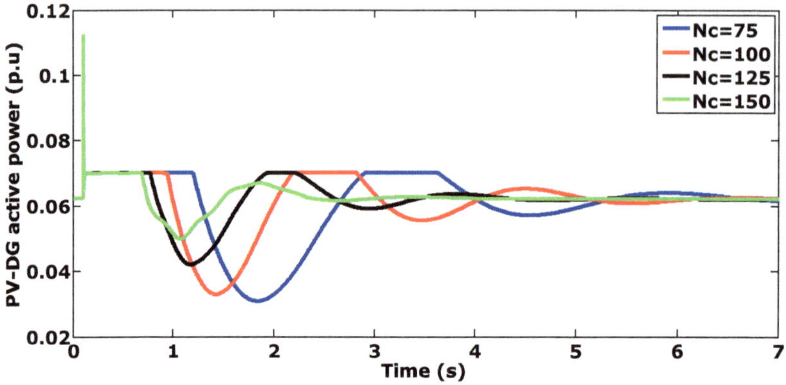

(d) PV-DG active power during the worst-case load disturbance

Fig. 5.3 (continued)

the sudden load increase, SG-DG releases some of its stored kinetic energy in the rotor. As shown in Fig. 5.4a, the rotor speed drops from its nominal value of 377 rad/s triggering the LTI-MPC to change the turbine reference set-point (P_{ref}) optimally. As a result, as shown in Fig. 5.4b, the SG-DG's active power starts increasing to accommodate the load change and to bring the rotor speed back to its nominal value. Since there is a primary reserve margin of 15% available for PV-DG, it follows an artificial droop characteristic assigned by the MPC to participate in the frequency control during the transient period. As shown in Fig. 5.4c, the PV-DG's active power (P_{G3}) followed droop characteristics as defined in (5.13) during the transient period and restored to its nominal value of 0.2154 p.u (85% of the MPP) at the steady-state. Thus in the steady-state, the complete active power load change and the associated network loss are accommodated by the SG-DG, whereas the PV-DG participated in the transient frequency control and has nothing to do with steady-state load change accommodation.

The SG-DG's active power has increased from an initial value of 0.4767 p.u to a final value of 0.5203 p.u. The frequency has deviated to a minimum of 376.9783 rad/s. It is restored to its nominal value of 377 rad/s at nearly 2.4 s, thus taking a restoration time of 2.3 s. To control and restore the generator bus voltages, the LTI-MPC optimally adjusts the voltage regulator input of the SG-DG and the $d3$–$q3$ axis components of the inverter output voltage. As a result, as shown in Fig. 5.4d, e, there is an optimal reactive power adjustment between the generators. The optimal adjustment is made in such a way that there will be reactive power balance in the network and voltage control at the generator buses happening all the time. As shown in Fig. 5.4f, g, the voltages at the generator buses B1 and B3 are restored to 1 p.u at approximately 0.35 s, thus taking a restoration time of 0.25 s. The voltage at bus B1 deviates to a minimum of 0.9892 p.u. The voltage at bus B3 deviates to a minimum of 0.9837 p.u. All the bus voltages and the system frequency are within limits during the transient and the steady-state periods as per the IEEE 1547.4 standard.

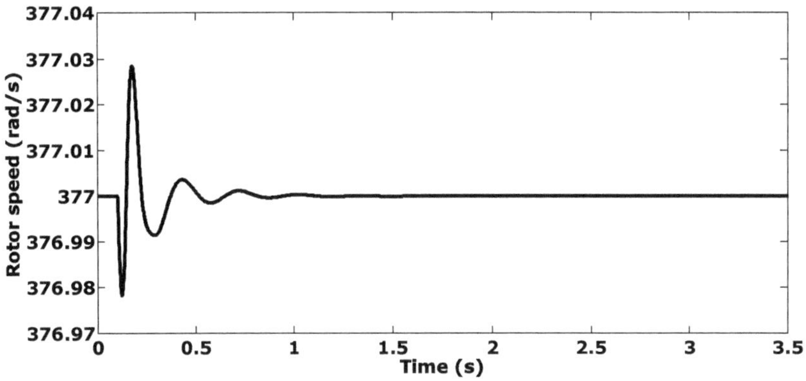

(a) Rotor speed during the single R-L load disturbance at bus B7

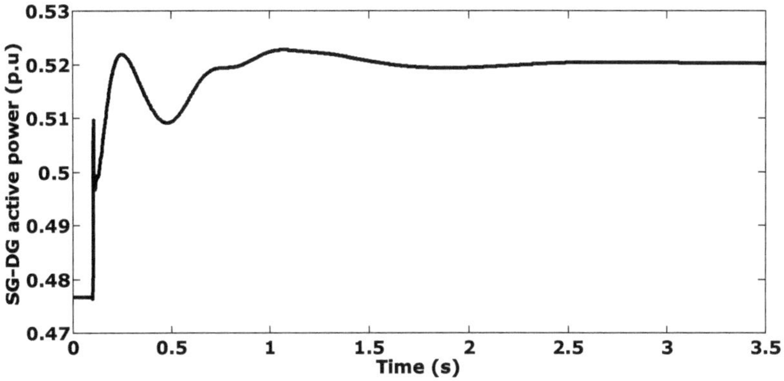

(b) SG-DG active power during the single R-L load disturbance at bus B7

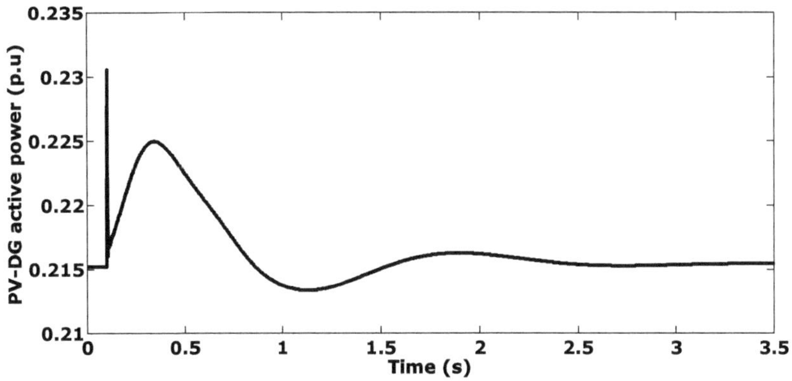

(c) PV-DG active power during the single R-L load disturbance at bus B7

Fig. 5.4 LTI-MPC performance for single R-L load disturbance at bus B7: **a** rotor speed, **b** SG-DG active power, **c** PV-DG active power, **d** SG-DG reactive power, **e** PV-DG reactive power, **f** bus B1 voltage, and **g** bus B3 voltage

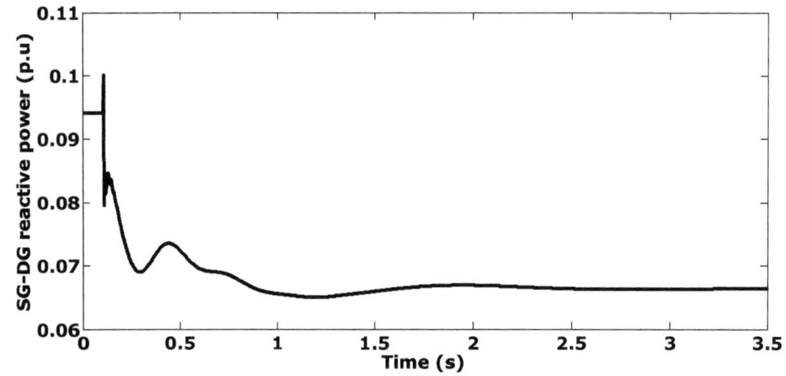

(d) SG-DG reactive power during the single R-L load disturbance at bus B7

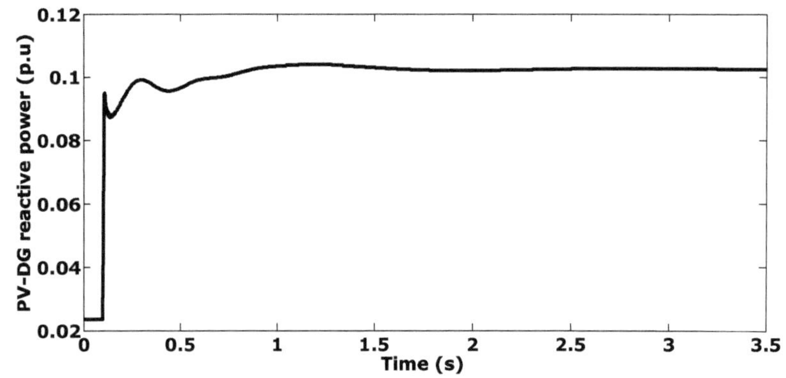

(e) PV-DG reactive power during the single R-L load disturbance at bus B7

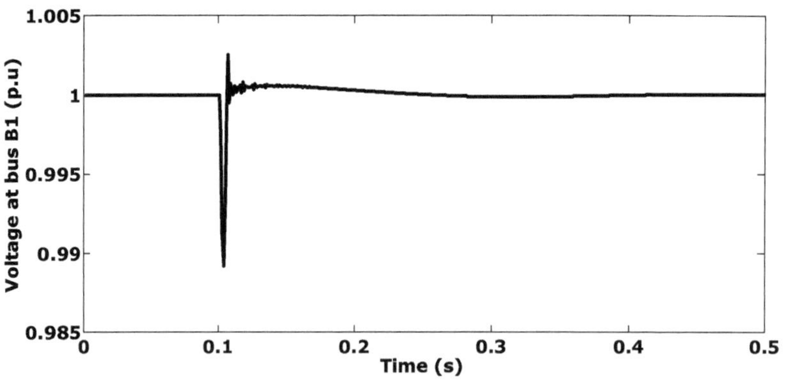

(f) Bus B1 voltage during the single R-L load disturbance at bus B7

Fig. 5.4 (continued)

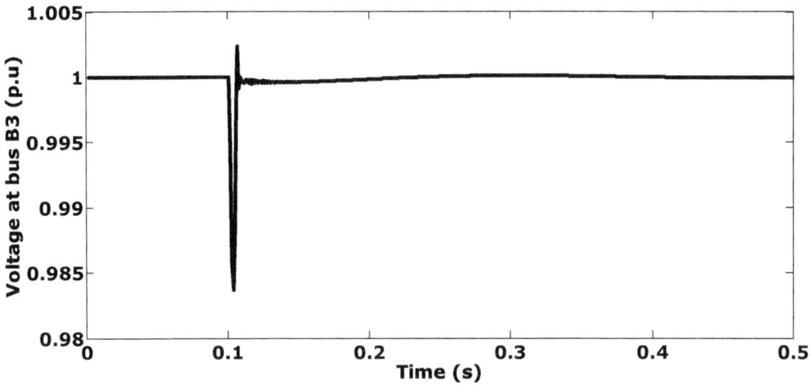

(g) Bus B3 voltage during the single R-L load disturbance at bus B7

Fig. 5.4 (continued)

Example 5.2: Multiple R-L and IM load disturbances at a time

This is a case to represent the large-signal load disturbance in the micro-grid. At $t = 0.1$ s, the p.u R-L loads L5, L7, L8 at buses B5, B7, B8 are suddenly changed all at a time from their initial values of 6.0941 + j0.0849, 3.8306 + j3.8757, 3.8079 + j3.8527 to their maximum possible values of 4.5388 + j0.0640, 2.7948 + j2.8276, 2.7948 + j2.8277. At the same time, the induction motor load torque at bus B6 (T_{mm6}) is changed to its maximum value of 0.1940 p.u, from an initial value of 0.1440 p.u. Due to the sudden load increase at multiple buses at a time, there is a considerable rotor speed drop, as shown in Fig. 5.5a. To control and restore the rotor speed to its nominal value, LTI-MPC increases the SG-DG's active power by optimally controlling its turbine reference set-point. At the same time, LTI-MPC allows the PV-DG's active power to follow an artificial droop characteristic during the transient period. However, in Fig. 5.5c, we can observe that at $t = 0.23$ s, the PV-DG's output has reached its maximum allowable limit of 0.2430 p.u, which is 96% of the PV array's maximum output. The 4% margin is to account for PV-DG losses. At this stage, the maximum power constraint of PV-DG explained in Sect. 5.2.2 becomes active. Thus from $t = 0.23$ s to $t = 0.42$ s, PV-DG could not participate in frequency regulation as the reference power generated from its droop characteristic is more than its maximum power. After $t = 0.42$ s, it again starts participating in the frequency regulation as its reference value is below the maximum power limit.

At the steady-state, the PV-DG's active power restored to its nominal value of 0.2154 p.u, which is 85% of its maximum power. The SG-DG's active power has increased from an initial value of 0.4767 p.u to a final value of 0.6706 p.u, as shown in Fig. 5.5b. It accounts for the complete steady-state load change and the corresponding network loss. The frequency has deviated to a minimum of 376.9057 rad/s and restored at nearly 3.4 s, taking a restoration time of 3.3 s. Figure 5.5d, e show the reactive power outputs of the SG-DG and PV-DG, respectively. The reactive powers

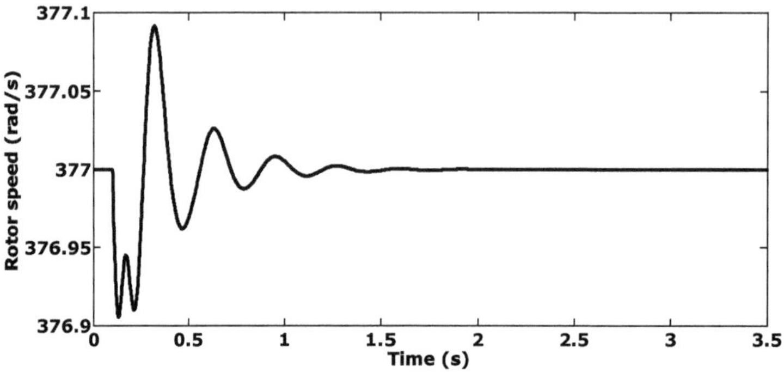

(a) Rotor speed during the multiple R-L and IM load disturbances at a time

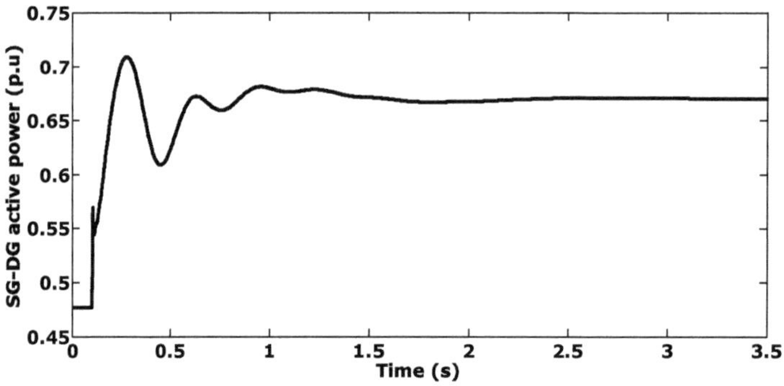

(b) SG-DG active power during the multiple R-L and IM load disturbances at a time

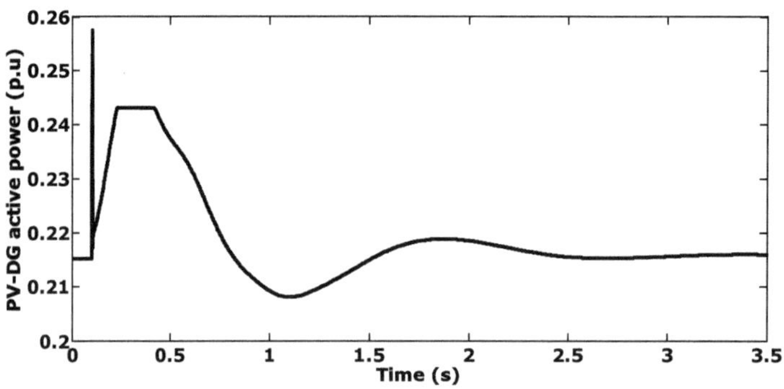

(c) PV-DG active power during the multiple R-L and IM load disturbances at a time

Fig. 5.5 LTI-MPC performance for multiple R-L and IM load disturbances at a time: **a** rotor speed, **b** SG-DG active power, **c** PV-DG active power, **d** SG-DG reactive power, **e** PV-DG reactive power, **f** bus B1 voltage, and **g** bus B3 voltage

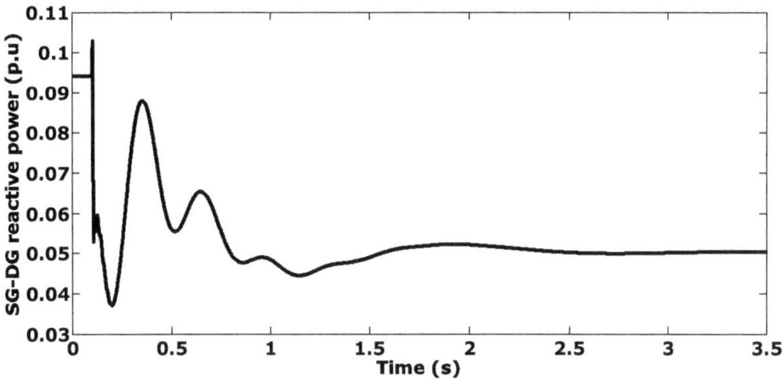

(d) SG-DG reactive power during the multiple R-L and IM load disturbances at a time

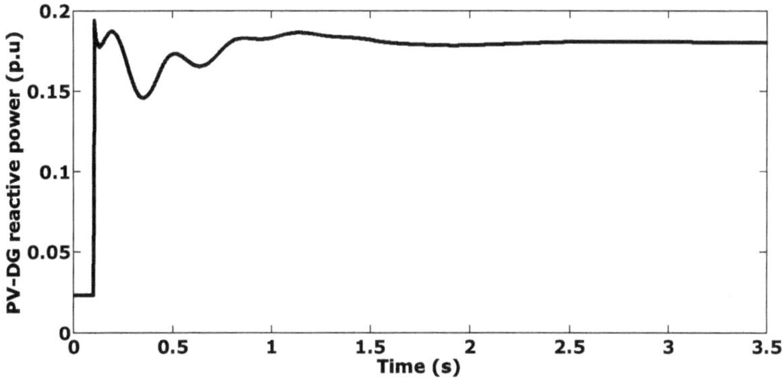

(e) PV-DG reactive power during the multiple R-L and IM load disturbances at a time

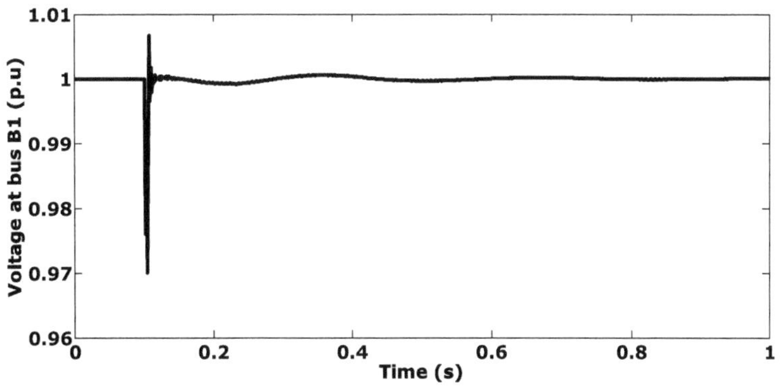

(f) Bus B1 voltage during the multiple R-L and IM load disturbances at a time

Fig. 5.5 (continued)

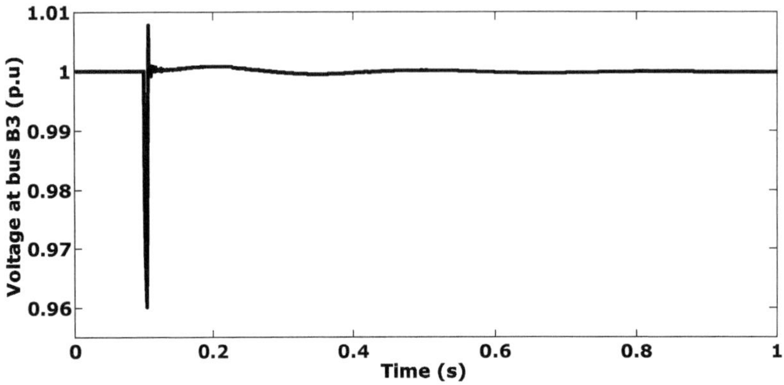

(g) Bus B3 voltage during the multiple R-L and IM load disturbances at a time

Fig. 5.5 (continued)

are optimally adjusted among the generators so that simultaneous reactive power balance in the network and voltage control at the generator buses happens. The voltages at the generator buses B1 and B3 are shown in Fig. 5.5f, g. The voltage at B1 deviates to a minimum of 0.97 p.u, and voltage at B3 deviates to a minimum of 0.96 p.u. They are restored to 1 p.u at approximately 0.75 s, taking a restoration time of 0.65 s.

Example 5.3: PV-DG source intermittency

In this case, at $t = 0.1$ s, the solar irradiance is suddenly changed from 800 to 650 W/m^2. Change in the irradiance causes the maximum output power of the PV array to drop from 0.2532 to 0.2032 p.u. Thus the new nominal power of the PV-DG at steady-state should be 85% of 0.2032 p.u, which is 0.1727 p.u. The steady-state nominal power change of PV-DG is seen as a disturbance in the micro-grid by the LTI-MPC. The performance curves for the source intermittency are shown in Fig. 5.6. It is evident from Fig. 5.6c that the PV-DG has reached the new steady-state power of 0.1727 p.u at 2.4 s. Figure 5.6b shows the active power output of the SG-DG. Since the load in the micro-grid is same, the steady-state value of the SG-DG active power has increased from 0.4767 to 0.5195 p.u to compensate for the decrease in the PV-DG's output. Figure 5.6d and e show the reactive power outputs of the SG-DG and PV-DG, respectively. The reactive powers are optimally adjusted among the generators so that simultaneous reactive power balance in the network and voltage control at the generator buses happens. The system frequency and voltages at the generator buses are within the limits during the transient period and restored to their nominal values during the steady-state. Thus we can conclude that the LTI-MPC can adequately handle the source intermittency.

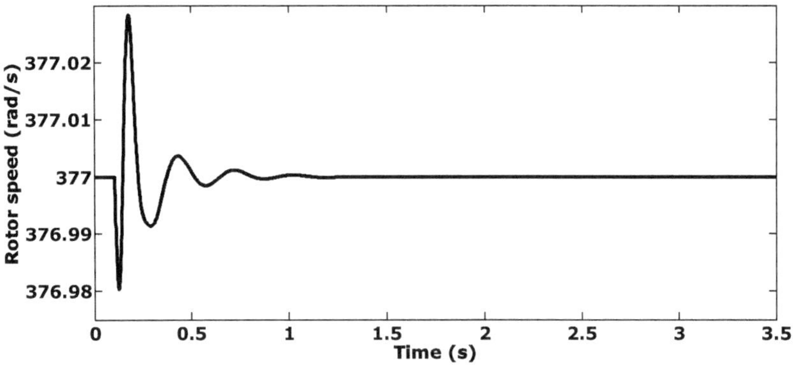

(a) Rotor speed during the PV-DG source intermittency

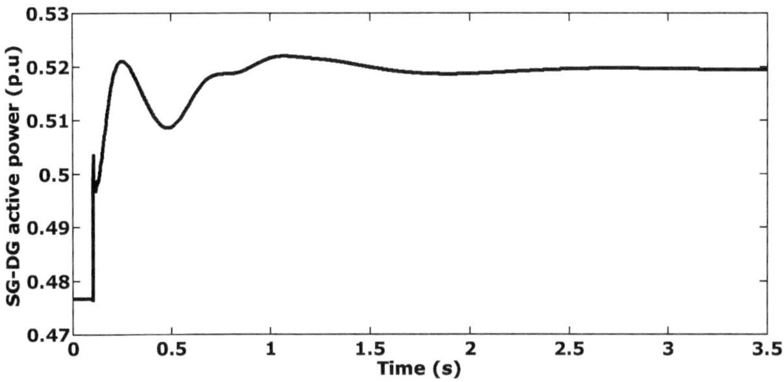

(b) SG-DG active power during the PV-DG source intermittency

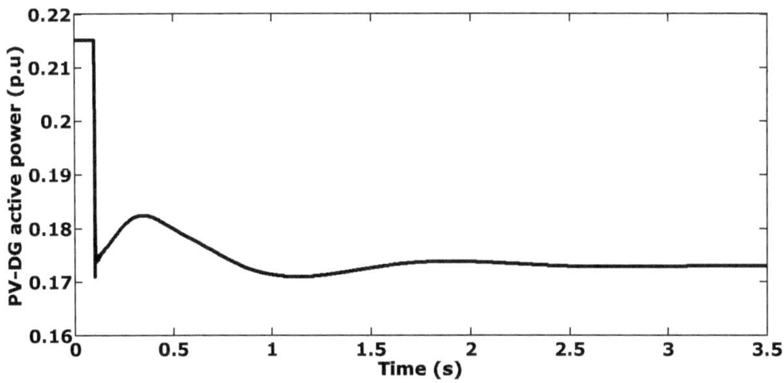

(c) PV-DG active power during the PV-DG source intermittency

Fig. 5.6 LTI-MPC performance for the PV-DG source intermittency: **a** rotor speed, **b** SG-DG active power, **c** PV-DG active power, **d** SG-DG reactive power, **e** PV-DG reactive power, **f** bus B1 voltage, and **g** bus B3 voltage

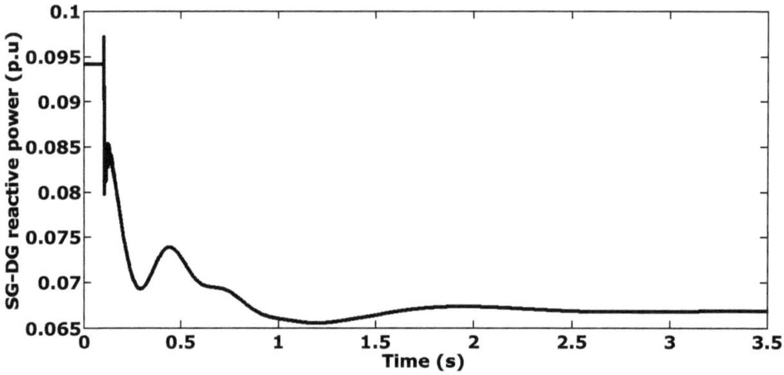

(d) SG-DG reactive power during the PV-DG source intermittency

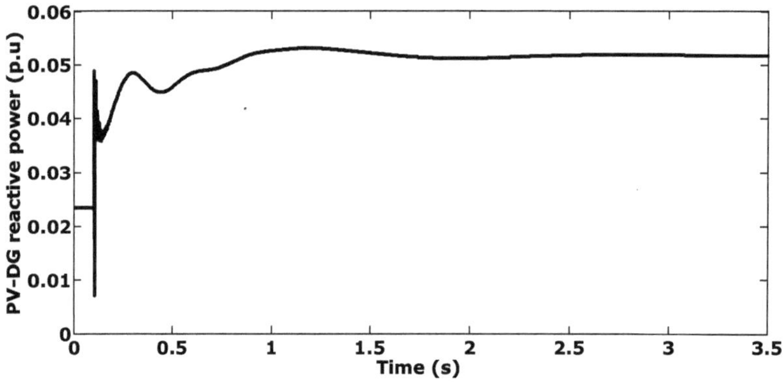

(e) PV-DG reactive power during the PV-DG source intermittency

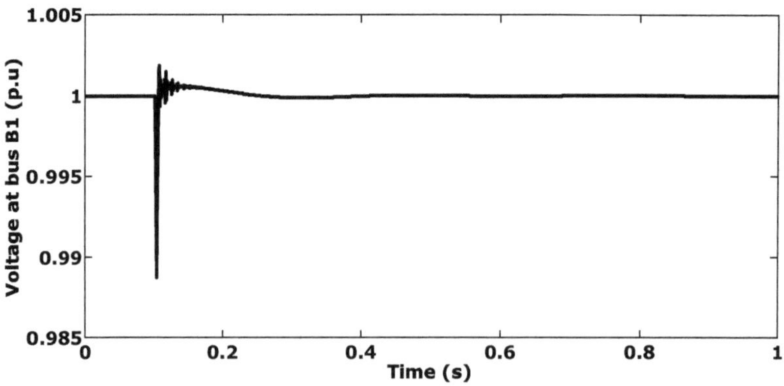

(f) Bus B1 voltage during the PV-DG source intermittency

Fig. 5.6 (continued)

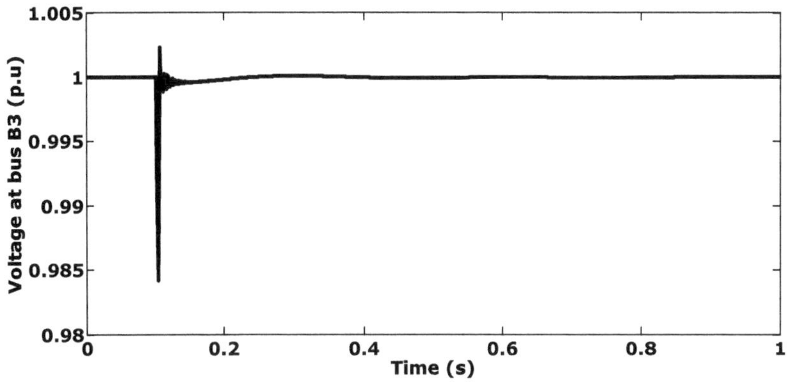

(g) Bus B3 voltage during the PV-DG source intermittency

Fig. 5.6 (continued)

5.4 Key Takeaways

- The chapter discussed the detailed mathematical formulation of the LTI-MPC for the micro-grid control.
- The LTI model of the micro-grid can be obtained by the linearization of the nonlinear micro-grid model around an operating point corresponding to each sample.
- The LTI model is used for the prediction of the micro-grid behaviour within the prediction horizon.
- The simulation results show that the LTI-MPC can adequately handle different kinds of disturbances in the micro-grid.
- The LTI-MPC formulation is very simple and involves fewer computations.
- The predicted behaviour can be directly expressed using closed-form equations.
- It can be very easily implemented online and is one of the brightest prospects for the centralized primary control of micro-grids.

References

Blet N, Megias D, Serrano J, De Prada C (2002) Nonlinear MPC versus MPC using on-line linearisation—a comparative study. IFAC Proc Vol 35(1):147–152

Ławryńczuk M (2015) Nonlinear state-space predictive control with on-line linearisation and state estimation. Int J Appl Math Comput Sci 25(4):833–847

Megias D, Serrano J, El Ghoumari MY (1999) Extended linearised predictive control: Practical control algorithms for non-linear systems. In: 1999 European control conference (ECC). IEEE, pp 2707–2712

Ranga VPSRV, Sesha S, Kesanakurthy SS (2018) Model predictive control approach for frequency and voltage control of standalone micro-grid. IET Gener Transm Distrib 12(14):3405–3413

Sagar PSV, Swarup KS (2016) Load frequency control in isolated micro-grids using centralized model predictive control. In: 2016 IEEE international conference on power electronics, drives and energy systems (PEDES). IEEE, pp 1–6

Vidyasagar PS, Swarup KS (2016) Discrete model predictive frequency and voltage control of isolated micro-grid in smart grid scenario. In: 2016 National power systems conference (NPSC). IEEE, pp 1–6

Vidyasagar PS, Swarup KS (2018) Voltage and frequency control of isolated micro-grid using extended linear model predictive controller. In: 2018 IEEE Power and Energy Society General Meeting (PESGM), IEEE, pp 1–5

Chapter 6
LTV-MPC with Extended "TAIL"

Abstract MPC with LTI approximation of the nonlinear model of the micro-grid is fairly simple and easy to implement. However, the LTI model is generally inaccurate and closely approximates the nonlinear model only in the small neighbourhood of the operating point around which the linearization is performed. The LTI model deviates from the original nonlinear model with an increase in the prediction horizon length N_p. Hence the LTI-MPC is limited to small prediction horizons. When a load disturbance or source intermittency occurs in the micro-grid, LTI-MPC indeed controls the micro-grid and moves it towards the new steady-state. However, it leads to large oscillatory transient response and large settling times due to the highly inaccurate prediction. Due to the receding horizon principle, control inputs that correspond to the present sample are chosen from the optimal control trajectories generated at each sample. The rest of the trajectories ("tail") are neglected. This is a waste of information that is optimal in some sense. A linear time-variant MPC (LTV-MPC) with an extended "tail" for the primary control of micro-grids is discussed in this chapter to tackle the difficulties mentioned above. The inspiration for the LTV-MPC is drawn from Kouvaritakis et al. (Int J Control 72:919–928, 1999) given at the end of the chapter. Puvvula et al. (IET Renew Power Gener 14:2221–2231, 2020) gives a detailed insight into different aspects of the LTV-MPC which forms the crux of this chapter.

Keywords Forced response · Linear time-variant approximation · Natural response · Optimal "tail"

6.1 Mathematical Formulation of the LTV-MPC

At each sample, within the prediction horizon, the nonlinear model of the micro-grid is linearized around the state and input reference trajectories. The linearization results in an LTV model of the micro-grid. The micro-grid response within the prediction horizon due to the input reference trajectories is referred to as the "Natural response". This response is calculated by directly solving the nonlinear mathematical model of the micro-grid along the input reference trajectories. The micro-grid response within

P. Vidyasagar and K. Shanti Swarup, *Design and Development of Model Predictive Primary Control of Micro Grids*, Springer Tracts in Electrical and Electronics Engineering, https://doi.org/10.1007/978-981-19-5852-6_6

the prediction horizon due to the perturbed inputs around the reference input trajectories is referred to as "Forced response". The perturbed inputs are coming from the MPC. This response is calculated using the LTV model of the micro-grid, which is obtained from the approximation of the nonlinear model around the reference input and state trajectories. The sum of the forced and natural response forms the complete predicted response of the micro-grid. This predicted behaviour is used in the formulation of the objective function of the LTV-MPC, which is quadratic. The quadratic objective function is evaluated subjected to a set of micro-grid operational constraints, thus generating the optimal control trajectories within the control horizon. Due to the receding horizon principle, inputs corresponding to the present time instant in the optimal control trajectories are applied to the micro-grid actuators. The rest of the optimal control trajectories are neglected. These neglected and unused trajectories form the "tail" of the LTV-MPC, which has optimal or suboptimal information. This "tail" computed in the present prediction horizon forms the input reference trajectories for the next prediction horizon when the following sample arrives.

6.1.1 Prediction of the Forced Response

The nonlinear model is given by the following continuous-time state-space model:

$$\frac{\mathrm{d}\boldsymbol{X}(t)}{\mathrm{d}t} = \boldsymbol{f}(\boldsymbol{X}(t), \boldsymbol{V}(t))$$

$$\boldsymbol{Y}(t) = \boldsymbol{g}(\boldsymbol{X}(t)) \tag{6.1}$$

$\boldsymbol{X}$ is the continuous-time state vector with n states, $\boldsymbol{V}$ is the input vector with n_{ip} inputs, and $\boldsymbol{Y}$ is the output vector with n_{op} outputs. Let $t = t_s$ seconds be the present continuous-time moment whose discrete-time equivalent is given by $k = k_i$. T_s is the sampling time of the controller. N_p and N_c are the lengths of the prediction horizon and control horizon, respectively. The prediction horizon is given by $\left(t_s \le t \le t_s + N_p T_s\right)$ in the continuous-time domain and in the discrete-time domain is given by $\left(k_i \le k \le k_i + N_p\right)$. The control horizon is given by $(t_s \le t \le t_s + (N_c - 1)T_s)$ in continuous-time domain, and in the discrete-time domain, control horizon is given by $(k_i \le k \le k_i + N_c - 1)$. Within the prediction horizon, let us consider the following:

$$\boldsymbol{X}(t) = \boldsymbol{X}_{\mathbf{ref}}(t) + \Delta \boldsymbol{X}(t), \quad \left(t_s \le t \le t_s + N_p T_s\right) \tag{6.2}$$

$$\boldsymbol{V}(t) = \boldsymbol{V}_{\mathbf{ref}}(t) + \Delta \boldsymbol{V}(t), \quad \left(t_s \le t \le t_s + N_p T_s\right) \tag{6.3}$$

$$\boldsymbol{Y}(t) = \boldsymbol{Y}_{\mathbf{ref}}(t) + \Delta \boldsymbol{Y}(t), \quad \left(t_s \le t \le t_s + N_p T_s\right) \tag{6.4}$$

$\boldsymbol{X}_{\mathbf{ref}}$, $\boldsymbol{V}_{\mathbf{ref}}$, and $\boldsymbol{Y}_{\mathbf{ref}}$ are the reference trajectories of the state vector, input vector, and output vector of the micro-grid. $\Delta\boldsymbol{X}$, $\Delta\boldsymbol{V}$, and $\Delta\boldsymbol{Y}$ are the perturbations around their respective reference trajectories. Substituting (6.2), (6.3), and (6.4) in the nonlinear model (6.1) and linearizing it around the reference trajectories using Taylor's theorem, we get the following linear time-variant (LTV) model:

$$\frac{\mathrm{d}\boldsymbol{Z}_c(t)}{\mathrm{d}t} = \boldsymbol{A}_c(t)\boldsymbol{Z}_c(t) + \boldsymbol{B}_c(t)\boldsymbol{u}(t) \tag{6.5}$$

$$\Delta\boldsymbol{Y}(t) = \boldsymbol{C}_c(t)\boldsymbol{Z}_c(t), \quad \left(t_s \le t \le t_s + N_p T_s\right)$$

$$\boldsymbol{Z}_c(t) = \Delta\boldsymbol{X}(t), \boldsymbol{u}(t) = \Delta\boldsymbol{V}(t), \boldsymbol{A}_c(t) = \left.\frac{\partial \boldsymbol{f}}{\partial \boldsymbol{X}}\right|_{\substack{X(t) = \boldsymbol{X}_{\mathbf{ref}}(t) \\ V(t) = \boldsymbol{V}_{\mathbf{ref}}(t)}},$$

$$\boldsymbol{B}_c(t) = \left.\frac{\partial \boldsymbol{f}}{\partial \boldsymbol{V}}\right|_{\substack{X(t) = \boldsymbol{X}_{\mathbf{ref}}(t) \\ V(t) = \boldsymbol{V}_{\mathbf{ref}}(t)}}, \boldsymbol{C}_c(t) = \left.\frac{\partial \boldsymbol{g}}{\partial \boldsymbol{X}}\right|_{\substack{X(t) = \boldsymbol{X}_{\mathbf{ref}}(t) \\ V(t) = \boldsymbol{V}_{\mathbf{ref}}(t)}}$$

The LTV model (6.5) can be discretized using the zero-order hold (ZOH) with a sampling time of T_s seconds, which is given by:

$$\boldsymbol{Z}_c(k+1) = \boldsymbol{A}_d(k)\boldsymbol{Z}_c(k) + \boldsymbol{B}_d(k)\boldsymbol{u}(k)$$

$$\Delta\boldsymbol{Y}(k) = \boldsymbol{C}_d(k)\boldsymbol{Z}_c(k), \quad \left(k_i \le k \le k_i + N_p\right) \tag{6.6}$$

To incorporate the integral action into the LTV-MPC, the LTV model (6.6) is augmented, resulting in a new state vector $\boldsymbol{Z}(k)$:

$$\boldsymbol{Z}(k+1) = \boldsymbol{A}(k)\boldsymbol{Z}(k) + \boldsymbol{B}(k)\Delta\boldsymbol{u}(k)$$

$$\Delta\boldsymbol{Y}(k) = \boldsymbol{C}(k)\boldsymbol{Z}(k), \quad \left(k_i \le k \le k_i + N_p\right) \tag{6.7}$$

$$\boldsymbol{A} = \begin{bmatrix} \boldsymbol{A}_d & \boldsymbol{B}_d \\ \boldsymbol{0}_{n_{\mathrm{ip}} \times n} & \boldsymbol{I}_{n_{\mathrm{ip}} \times n_{\mathrm{ip}}} \end{bmatrix}, \boldsymbol{B} = \begin{bmatrix} \boldsymbol{B}_d \\ \boldsymbol{I}_{n_{\mathrm{ip}} \times n_{\mathrm{ip}}} \end{bmatrix}, \boldsymbol{C} = \begin{bmatrix} \boldsymbol{C}_d & \boldsymbol{0}_{n_{\mathrm{op}} \times n_{\mathrm{ip}}} \end{bmatrix},$$

$$\boldsymbol{Z}(k) = \begin{bmatrix} \boldsymbol{Z}_c(k)^{\mathrm{T}} & \boldsymbol{u}(k-1)^{\mathrm{T}} \end{bmatrix}^{\mathrm{T}}, \Delta\boldsymbol{u}(k) = \boldsymbol{u}(k) - \boldsymbol{u}(k-1)$$

Let $\Delta\boldsymbol{Y}_p$ be the vector that contains the predicted values of the forced response of the micro-grid within the prediction horizon:

$$\Delta\boldsymbol{Y}_p = \left[\Delta\boldsymbol{Y}(k_i+1)^{\mathrm{T}} \; \Delta\boldsymbol{Y}(k_i+2)^{\mathrm{T}} \ldots \Delta\boldsymbol{Y}\left(k_i+N_p\right)^{\mathrm{T}}\right]^{\mathrm{T}} \tag{6.8}$$

Let $\boldsymbol{U}$ be the vector that contains the future control trajectories within the control horizon:

$$\boldsymbol{U} = \left[\, \Delta\boldsymbol{u}(k_i)^{\mathrm{T}}\ \Delta\boldsymbol{u}(k_i+1)^{\mathrm{T}}\ \ldots\ \Delta\boldsymbol{u}(k_i+N_c-1)^{\mathrm{T}} \,\right]^{\mathrm{T}} \tag{6.9}$$

From (6.7), we can write $\Delta\boldsymbol{Y}(k_i+m)$ as:

$$\Delta\boldsymbol{Y}(k_i+m) = \boldsymbol{C}(k_i+m)\boldsymbol{Z}(k_i+m)$$

Also, from (6.7), we can write $\boldsymbol{Z}(k_i+m)$ as:

$$\begin{aligned}\boldsymbol{Z}(k_i+m) = &\left[\prod_{j=0}^{m-1} \boldsymbol{A}(k_i+m-1-j)\right]\boldsymbol{Z}(k_i)\\ &+\sum_{j=1}^{m} \boldsymbol{A}_{j,m}\boldsymbol{B}(k_i+j-1)\Delta\boldsymbol{u}(k_i+j-1)\end{aligned} \tag{6.10}$$

$$\boldsymbol{A}_{j,m} = \begin{cases} \prod\limits_{l=1}^{m-j} \boldsymbol{A}(k_i+m-l) & \text{for } m > j \\ \boldsymbol{I} & \text{for } m = j \end{cases} \tag{6.11}$$

$\boldsymbol{Z}(k_i)$ is the augmented state vector at present time instant $k = k_i$, which is zero in this case. Using (6.9), (6.10), and (6.11), the predicted vector (6.8) of the forced response of the micro-grid can be expressed as:

$$\Delta\boldsymbol{Y}_p = \boldsymbol{\varphi}_1\boldsymbol{Z}(k_i) + \boldsymbol{\varphi}_2\boldsymbol{U} \tag{6.12}$$

6.1.2 *Prediction of the Natural Response*

The natural response of the micro-grid within the prediction horizon can be calculated by solving the nonlinear model (6.1) along the state and input reference trajectories.

$$\begin{aligned}&\boldsymbol{X}_{\mathbf{ref}}(t) = \boldsymbol{X}(t_s) + \int_{t_s}^{t} \boldsymbol{f}(\boldsymbol{X}(t), \boldsymbol{V}_{\mathbf{ref}}(t))\mathrm{d}t\\ &\boldsymbol{Y}_{\mathbf{ref}}(t) = \boldsymbol{g}(\boldsymbol{X}_{\mathbf{ref}}(t)), \quad \left(t_s \le t \le t_s + N_pT_s\right)\end{aligned} \tag{6.13}$$

The continuous-time natural response trajectories of the micro-grid can be discretized using ZOH. Let $\boldsymbol{Y}_p$ be the vector that contains the predicted values of the natural response of the micro-grid within the prediction horizon:

$$\boldsymbol{Y}_p = \left[\boldsymbol{Y}_{\mathbf{ref}}(k_i+1)^{\mathrm{T}}\ \boldsymbol{Y}_{\mathbf{ref}}(k_i+2)^{\mathrm{T}} \ldots \boldsymbol{Y}_{\mathbf{ref}}\left(k_i+N_p\right)^{\mathrm{T}} \right]^{\mathrm{T}} \tag{6.14}$$

6.1.3 Optimal Control Problem Formulation

The complete predicted response of the micro-grid within the prediction horizon is given by the sum of the natural and forced response of the micro-grid:

$$\boldsymbol{Y}_E = \boldsymbol{Y}_p + \Delta\boldsymbol{Y}_p = \boldsymbol{Y}_p + \boldsymbol{\varphi}_1\boldsymbol{Z}(k_i) + \boldsymbol{\varphi}_2\boldsymbol{U} \tag{6.15}$$

At each sampling instant, once $\boldsymbol{Y}_E$ is formed from (6.15), a quadratic objective function 'J' is formulated using $\boldsymbol{Y}_E$ and is given by:

$$J = (\boldsymbol{W} - \boldsymbol{Y}_E)^{\mathrm{T}}(\boldsymbol{W} - \boldsymbol{Y}_E) + \boldsymbol{U}^{\mathrm{T}}\boldsymbol{R}\boldsymbol{U} \tag{6.16}$$

$\boldsymbol{W}$ is a vector with reference set-points for the micro-grid outputs within the prediction horizon. $\boldsymbol{R}$ is a positive definite weight matrix on input increments. Substituting (6.15) in (6.16), the complete quadratic optimal control problem is given by:

$$\begin{aligned}
\text{Minimize}\quad & J = \boldsymbol{U}^{\mathrm{T}}\boldsymbol{E}\boldsymbol{U} + 2\boldsymbol{U}^{\mathrm{T}}(\boldsymbol{E}_1 + \boldsymbol{E}_2 - \boldsymbol{E}_3) + \textbf{Constant} \\
\text{S.T.}\quad & \boldsymbol{E} = \boldsymbol{\varphi}_2^{\mathrm{T}}\boldsymbol{\varphi}_2 + \boldsymbol{R},\ \boldsymbol{E}_1 = \boldsymbol{\varphi}_2^{\mathrm{T}}\boldsymbol{\varphi}_1\boldsymbol{Z}(k_i),\ \boldsymbol{E}_2 = \boldsymbol{\varphi}_2^{\mathrm{T}}\boldsymbol{Y}_p,\ \boldsymbol{E}_3 = \boldsymbol{\varphi}_2^{\mathrm{T}}\boldsymbol{W}, \\
& \boldsymbol{u}_{\min} \le \boldsymbol{u}(k_i) \le \boldsymbol{u}_{\max},\ \boldsymbol{V}_{\min} \le \boldsymbol{V}(k_i) \le \boldsymbol{V}_{\max},\ \Delta\boldsymbol{u}_{\min} \le \Delta\boldsymbol{u}(k_i) \le \Delta\boldsymbol{u}_{\max}, \\
& P_{G1,\min} \le P_{G1}(k_i+1) \le P_{G1,\max},\ Q_{G1,\min} \le Q_{G1}(k_i+1) \le Q_{G1,\max}, \\
& P_{G3,\min} \le P_{G3}(k_i+1) \le P_{G3,\max},\ Q_{G3,\min} \le Q_{G3}(k_i+1) \le Q_{G3,\max}, \\
& V_{\mathrm{mppt}}(G,T) \le V_{\mathrm{dc}}(k_i+1) \le V_{\mathrm{oc}}(G,T)
\end{aligned} \tag{6.17}$$

The pairs (P_{G1}, Q_{G1}), ($P_{G1,\max}$, $Q_{G1,\max}$), and ($P_{G1,\min}$, $Q_{G1,\min}$) are active and reactive power outputs of the SG-DG, their upper and lower limits. The pairs (P_{G3}, Q_{G3}), ($P_{G3,\max}$, $Q_{G3,\max}$), and ($P_{G3,\min}$, $Q_{G3,\min}$) are active and reactive power outputs of thc PV-DG, their upper and lower limits. $V_{\mathrm{mppt}}(G, T)$ and $V_{\mathrm{oc}}(G, T)$ are the output voltages of the PV array corresponding to the maximum power point and open circuit. The above quadratic optimal control problem (6.17) is evaluated (minimized) to generate an optimal coefficient vector $\boldsymbol{U} = \boldsymbol{U}_{\mathbf{opt}}$. In (6.9), the optimal input vector corresponding to the present sampling instant is given by $\Delta\boldsymbol{u}(k_i) = \Delta\boldsymbol{u}_{\mathbf{opt}}(k_i) = \Delta\boldsymbol{u}_{\mathbf{opt}}(0)$. From $\Delta\boldsymbol{u}_{\mathbf{opt}}(k_i)$, we can calculate the $\boldsymbol{u}_{\mathbf{opt}}(k_i)$ and $\boldsymbol{V}_{\mathbf{opt}}(k_i)$. Due to the receding horizon principle, the optimal inputs corresponding to the present time $\boldsymbol{V}_{\mathbf{opt}}(k_i) = \boldsymbol{V}_{\mathbf{opt}}(t_s)$ are applied to the micro-grid neglecting rest of the optimal control trajectories. When the next sample arrives, the whole procedure represented by Eqs. (6.1)–(6.17) is repeated. The complete operational procedure of the proposed LTV-MPC is shown in Fig. 6.1.

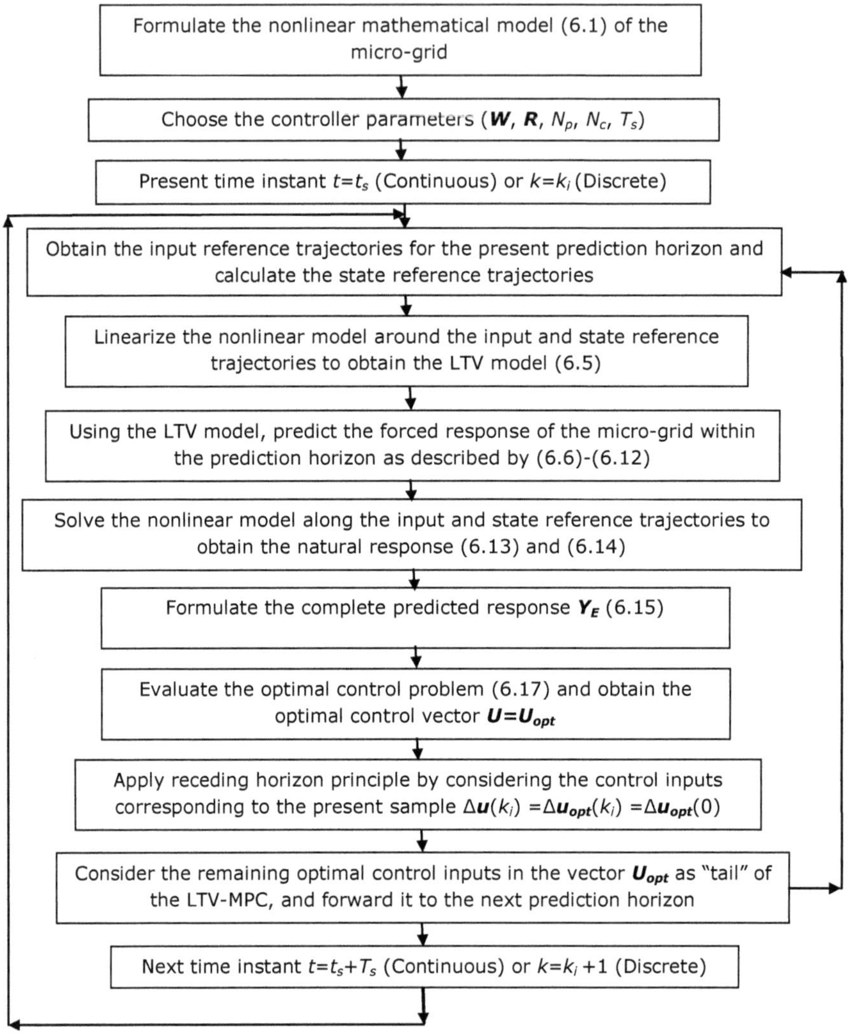

Fig. 6.1 Flowchart of the operation of the LTV-MPC

6.1.4 Choice of the Input Reference Trajectories $V_{\mathbf{ref}}(t)$

Let us consider a vector $\boldsymbol{V}_{\mathbf{tail}}$, which is the "tail" of the optimal control trajectories that were calculated and unused in the previous sampling instant ($k = k_i - 1$) due to the receding horizon strategy:

$$\boldsymbol{V}_{\mathbf{tail}} = \left[\, \boldsymbol{V}_{\mathbf{opt}}(k_i|k_i-1)^{\mathrm{T}} \; \boldsymbol{V}_{\mathbf{opt}}(k_i+1|k_i-1)^{\mathrm{T}} \ldots \boldsymbol{V}_{\mathbf{opt}}(k_i+N_c-1|k_i-1)^{\mathrm{T}} \,\right]^{\mathrm{T}} \tag{6.18}$$

The optimal inputs in the vector $\boldsymbol{V}_{\mathbf{tail}}$ are considered to be the input reference trajectories for the present control horizon ($k_i \leq k \leq k_i + N_c - 1$), , corresponding to the present sampling instant ($k = k_i$).

$$\boldsymbol{V}_{\mathbf{ref}}(k_i + m) = \boldsymbol{V}_{\mathbf{opt}}(k_i + m|k_i - 1), \quad (0 \leq m \leq N_c - 1) \tag{6.19}$$

Since a zero-order hold (ZOH) circuit is used for the discretization, we can write the input reference trajectories in the continuous-time domain as follows:

$$\boldsymbol{V}_{\mathbf{ref}}(t) = \boldsymbol{V}_{\mathbf{opt}}(k_i + m|k_i - 1), t_s + mT_s \leq t < t_s + (m+1)T_s \tag{6.20}$$

6.2 Performance Analysis

The upper limit of N_p in the case of LTV-MPC is higher than that of the LTI-MPC. It came around $N_p = 175$ for the same circumstances explained in Chap. 5, above which the simulation is taking a lot of time and getting stuck which is presumed due to numerical instability. However, the upper limit for N_p in the case of LTV-MPC is not so useful due to the following reasons:

- While choosing the N_p, we must consider the fact that the computational time of the control trajectories should be less than that of the sampling rate of the controller and there should be enough room for the communication delay.
- A prediction horizon length up to $N_p = 150$ satisfies the above requirements for LTI-MPC as described in Chap. 5.
- Hence we couldn't even cross the LTI-MPC upper limit for N_p.
- The LTV-MPC requires more computational time than that of the LTI-MPC for the same N_p.
- So even if we choose higher N_p values that produce feasible solutions in the case of LTV-MPC, it is difficult to implement them with rapid sampling rate requirements of the primary control level.
- Also, if we use N_p values greater than the upper limit of LTI-MPC, then the comparative analysis is not possible as the LTI-MPC becomes unstable for values higher than 164. Hence $N_p = N_c = 150$ is used for the analysis even in this chapter.

Example 6.1: Single R-L Load Disturbance
This is a case to represent the small-signal load disturbance in the micro-grid. At $t = 0.1$ s, the p.u-aggregated R-L load L7 at bus B7 ($R_{l7} + jX_{l7}$) of the micro-grid shown in Fig. 3.1 is suddenly changed from an initial value of $3.8306 + j3.8757$ to $2.7948 + j2.8276$. This change indicates a sudden load increase at bus B7. The performance of the LTV-MPC can be analysed using Fig. 6.2. To meet the sudden load increase, SG-DG releases some of its stored kinetic energy in the rotor. As shown in Fig. 6.2a, the rotor speed drops from its nominal value of 377 rad/s triggering the

LTV-MPC to change the turbine reference set-point (P_{ref}) optimally. As a result, as shown in Fig. 6.2b, the SG-DG's active power starts increasing to accommodate the load change and to bring the rotor speed back to its nominal value. Since there is a primary reserve margin of 15% available for PV-DG, it follows an artificial droop characteristic assigned by the MPC to participate in the frequency control during the transient period. As shown in Fig. 6.2c, the PV-DG's active power (P_{G3}) followed droop characteristics during the transient period and restored to its nominal value of 0.2154 p.u at the steady-state. Thus in the steady-state, the complete active power load change and the associated network loss are accommodated by the SG-DG, whereas the PV-DG participates in the transient frequency control and has nothing to do with the accommodation of the steady-state active power load change.

The SG-DG's active power has increased from an initial value of 0.4767 p.u to a final value of 0.5203 p.u. The frequency has deviated to a minimum of 376.9832 rad/s. It is restored to its nominal value of 377 rad/s at nearly 2.3 s, thus taking a restoration time of 2.2 s. To control and restore the generator bus voltages, the LTV-MPC optimally adjusts the voltage regulator input of SG-DG and the $d3$-$q3$ axis components of the inverter output voltage. As a result, as shown in Fig. 6.2d and e, there is an optimal reactive power adjustment between the generators. The optimal adjustment is made in such a way that there will be a reactive power balance and generator bus voltage control happening all the time. As shown in Fig. 6.2f and g, the voltages at the generator buses B1 and B3 are restored to 1 p.u at approximately 0.35 s, thus taking a restoration time of 0.25 s. The voltage at bus B1 deviates to a minimum of 0.9892 p.u. The voltage at bus B3 deviates to a minimum of 0.9837 p.u. All the bus voltages and the system frequency are within limits and adhere to IEEE 1547.4 standard.

Example 6.2: Multiple R-L Load and IM Load Disturbances at a Time

This is a case to represent the large-signal load disturbance in the micro-grid. At $t =$ 0.1 s, the p.u R-L loads L5, L7, and L8 at buses B5, B7, and B8 are suddenly changed all at a time from their initial values of $6.0941 + j0.0849$, $3.8306 + j3.8757$, and $3.8079 + j3.8527$ to their maximum possible values of $4.5388 + j0.0640$, $2.7948 + j2.8276$, and $2.7948 + j2.8277$. At the same time, the induction motor load torque at bus B6 (T_{mm6}) is changed to its maximum value of 0.1940 p.u, from an initial value of 0.1440 p.u. Due to the sudden load increase at multiple buses at a time, there is a considerable rotor speed drop, as shown in Fig. 6.3a. To control and restore the rotor speed to its nominal value, LTV-MPC increases the SG-DG's active power by optimally controlling its turbine reference set-point. At the same time, the LTV-MPC allows the PV-DG's active power to follow an artificial droop characteristic during the transient period. However, in Fig. 6.3c, we can observe that at $t = 0.23$ s, the PV-DG's output has reached its maximum allowable limit of 0.2430 p.u, which is 96% of the PV array's maximum output. The 4% margin is to account for PV-DG losses. At this stage, the maximum power constraint of PV-DG in (6.17) becomes active. Thus from $t = 0.23$ s to $t = 0.42$ s, PV-DG could not participate in frequency regulation as the reference power generated from its droop characteristic is more than

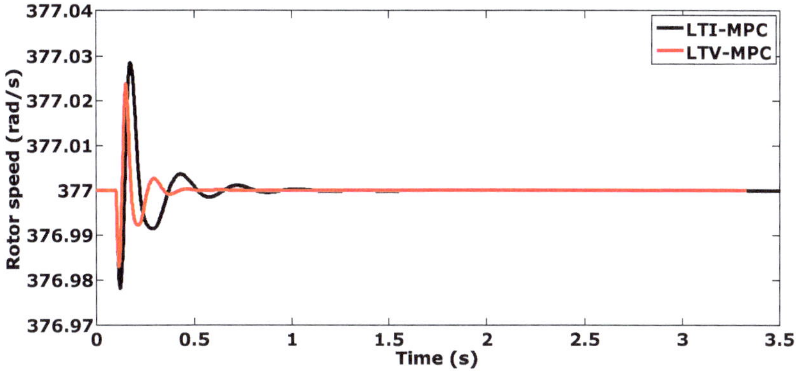

(a) Rotor speed during the single R-L load disturbance at bus B7

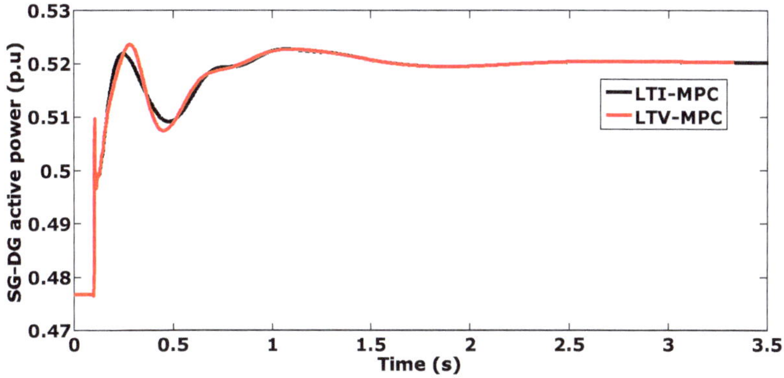

(b) SG-DG active power during the single R-L load disturbance at bus B7

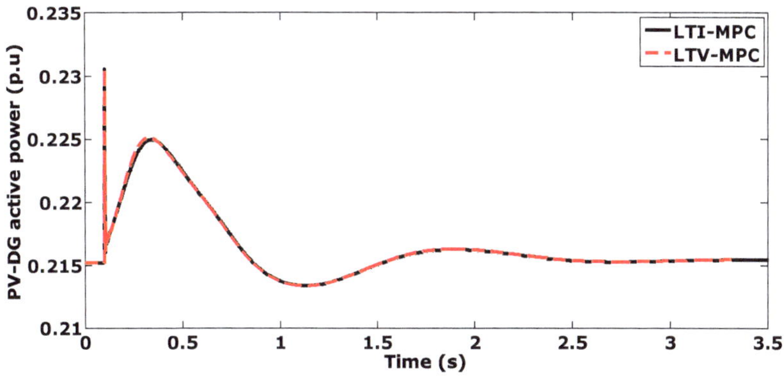

(c) PV-DG active power during the single R-L load disturbance at bus B7

Fig. 6.2 LTV-MPC performance for single R-L load disturbance at bus B7: **a** rotor speed, **b** SG-DG active power, **c** PV-DG active power, **d** SG-DG reactive power, **e** PV-DG reactive power, **f** bus B1 voltage, and **g** bus B3 voltage

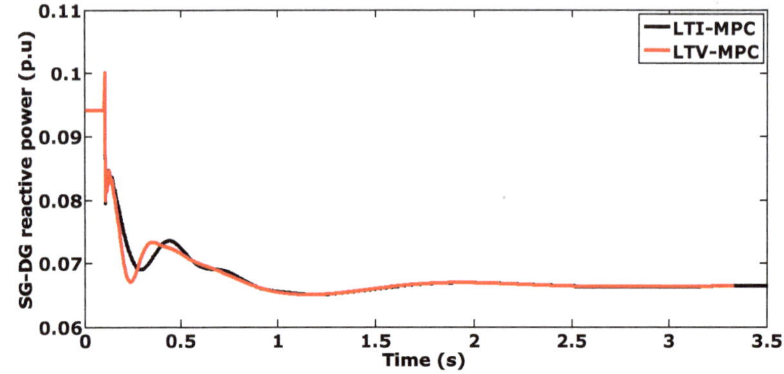

(d) SG-DG reactive power during the single R-L load disturbance at bus B7

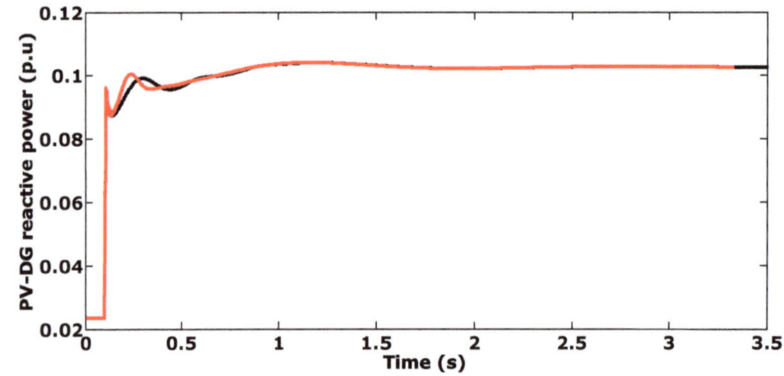

(e) PV-DG reactive power during the single R-L load disturbance at bus B7

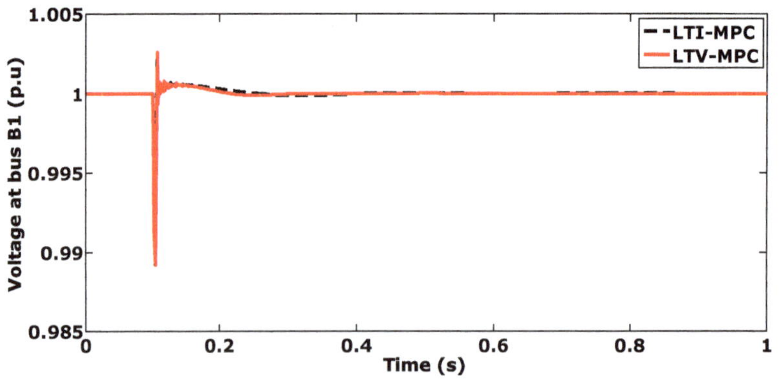

(f) Bus B1 voltage during the single R-L load disturbance at bus B7

Fig. 6.2 (continued)

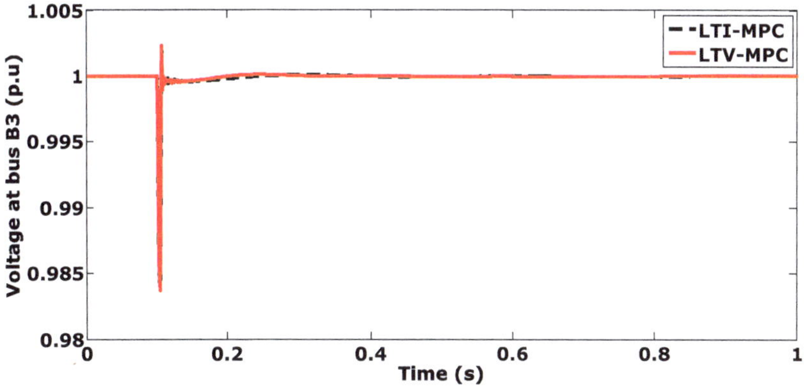

(g) Bus B3 voltage during the single R-L load disturbance at bus B7

Fig. 6.2 (continued)

its maximum power. After $t = 0.42$ s, it again starts participating in the frequency regulation.

At the steady-state, the PV-DG's active power is restored to its nominal value of 0.2154 p.u, which is 85% of its maximum power. The SG-DG's active power has increased from an initial value of 0.4767 p.u to a final value of 0.6708 p.u, as shown in Fig. 6.3b. It accounts for the complete steady-state load change and the corresponding network loss. The frequency has deviated to a minimum of 376.9226 rad/s and restored at nearly 3 s, taking a restoration time of 2.9 s. Figure 6.3d, e show the reactive power outputs of the SG-DG and PV-DG, respectively. The reactive powers are optimally adjusted among the generators so that simultaneous reactive power balance in the network and voltage control at the generator buses happens. The voltages at the generator buses B1 and B3 are shown in Fig. 6.3f, g. The voltage at bus B1 deviates to a minimum of 0.97 p.u, and voltage at B3 deviates to a minimum of 0.96 p.u. They are restored to 1 p.u at approximately 0.5 s, thus taking a restoration time of 0.4 s.

Example 6.3: PV-DG Source Intermittency

In this case, at $t = 0.1$ s, the solar irradiance is suddenly changed from 800 to 650 W/m^2. Change in the irradiance causes the maximum output power of the PV array to drop from 0.2532 p.u to 0.2032 p.u. Thus the new nominal power of PV-DG at steady-state should be 85% of 0.2032 p.u, which is 0.1727 p.u. The steady-state nominal power change of PV-DG is seen as a disturbance in the micro-grid by the LTV-MPC. The performance curves for the source intermittency are shown in Fig. 6.4. It is evident from Fig. 6.4c that the PV-DG has reached the new steady-state power of 0.1727 p.u at 2.3 s. Figure 6.4b shows the active power output of the SG-DG. Since the load in the micro-grid is the same, the steady-state value of the SG-DG active power has increased from 0.4767 p.u to 0.5195 p.u to compensate for the decrease in the PV-DG's output. Figure 6.4d, e show the reactive power outputs

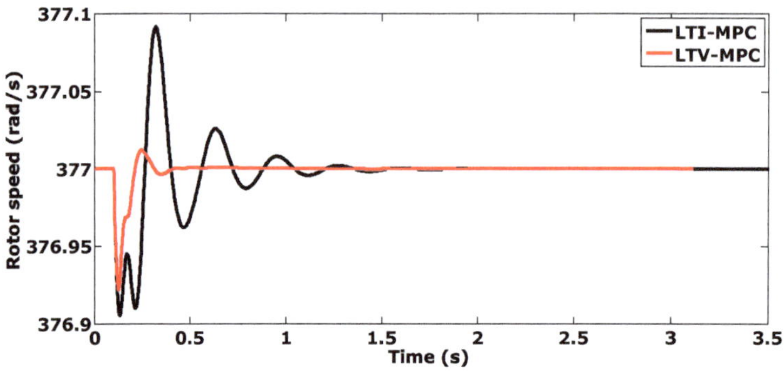

(a) Rotor speed during the multiple R-L and IM load disturbances at a time

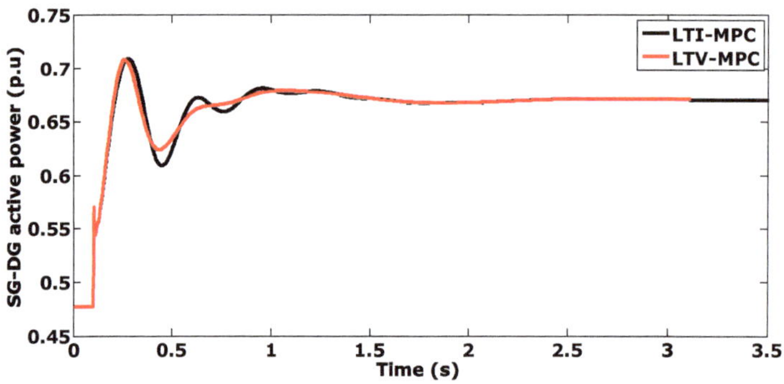

(b) SG-DG active power during the multiple R-L and IM load disturbances at a time

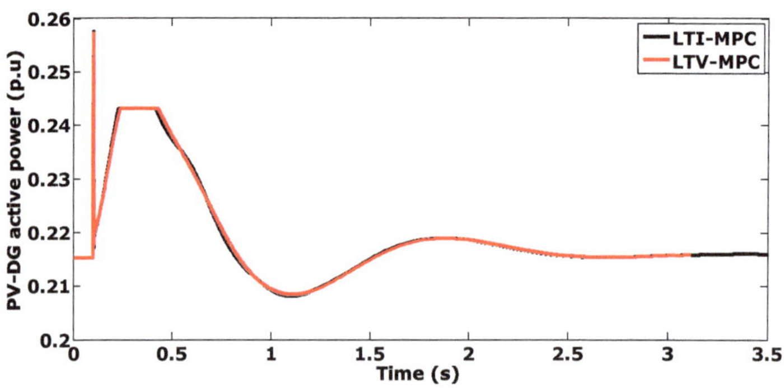

(c) PV-DG active power during the multiple R-L and IM load disturbances at a time

Fig. 6.3 LTV-MPC performance for multiple R-L and IM load disturbances at a time: **a** rotor speed, **b** SG-DG active power, **c** PV-DG active power, **d** SG-DG reactive power, **e** PV-DG reactive power, **f** bus B1 voltage, and **g** bus B3 voltage

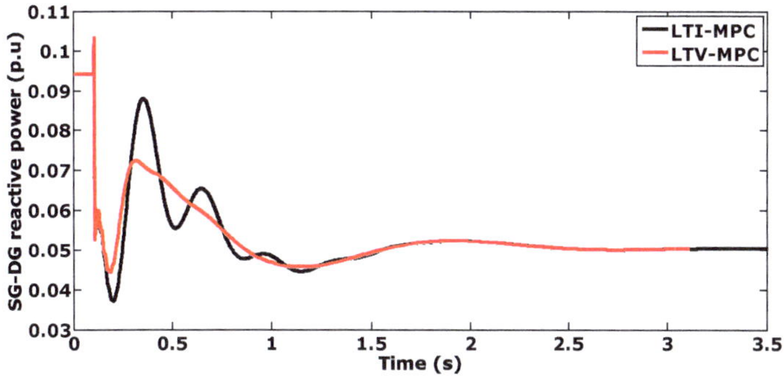

(d) SG-DG reactive power during the multiple R-L and IM load disturbances at a time

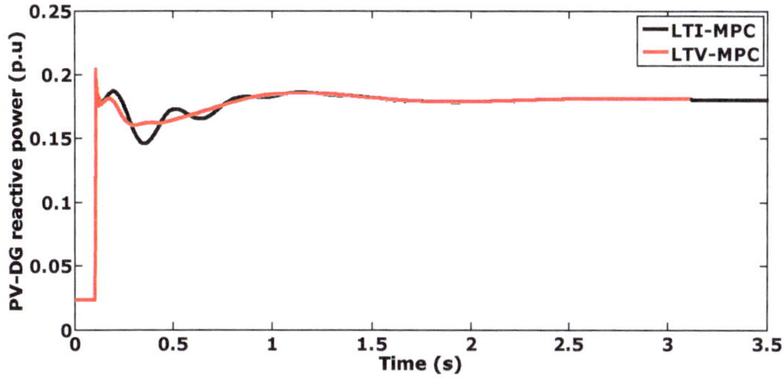

(e) PV-DG reactive power during the multiple R-L and IM load disturbances at a time

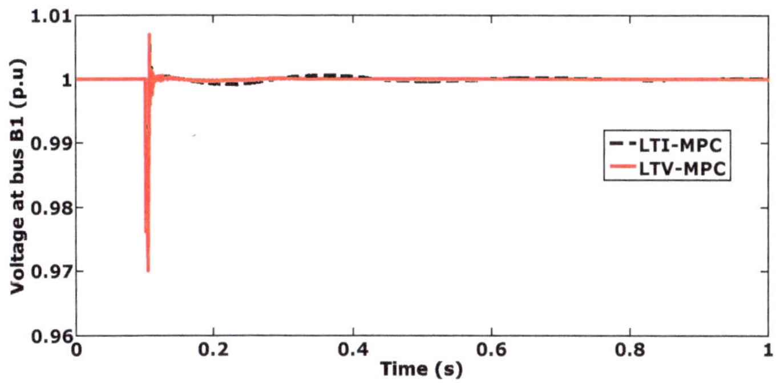

(f) Bus B1 voltage during the multiple R-L and IM load disturbances at a time

Fig. 6.3 (continued)

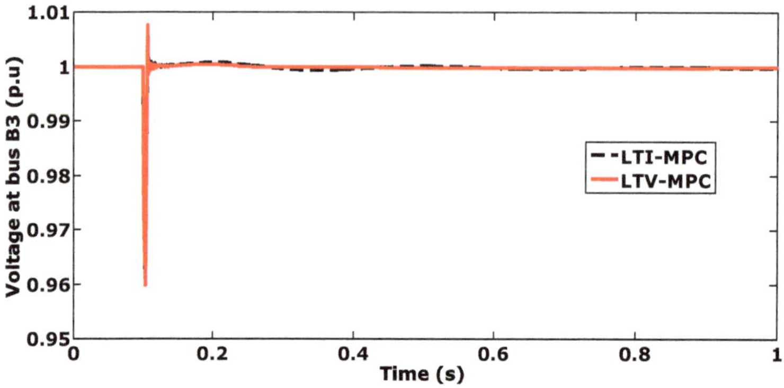

(g) Bus B3 voltage during the multiple R-L and IM load disturbances at a time

Fig. 6.3 (continued)

of the SG-DG and PV-DG, respectively. The reactive powers are optimally adjusted among the generators so that simultaneous reactive power balance in the network and voltage control at the generator buses happens. The system frequency and voltages at the generator buses are within limits during the transient period and restored to their nominal values during the steady-state.

6.3 Key Takeaways

- The chapter discussed the detailed mathematical formulation of the LTV-MPC for the micro-grid control.
- The LTV model of the micro-grid can be obtained by the linearization of the nonlinear micro-grid model around the state and input reference trajectories within the prediction horizon at each sample.
- The simulation results show that the LTV-MPC is superior in performance compared to the LTI-MPC. This superiority is due to the more accurate prediction of the micro-grid behaviour within the prediction horizon.
- Unlike LTI-MPC, the LTV-MPC allows large prediction horizons in its design.
- The chapter also discussed how to utilize the unused optimal "tail" calculated at each sample as the reference input trajectories for the next sample.
- The LTV-MPC involves more computations than the LTI-MPC at each sample but provides much better performance.

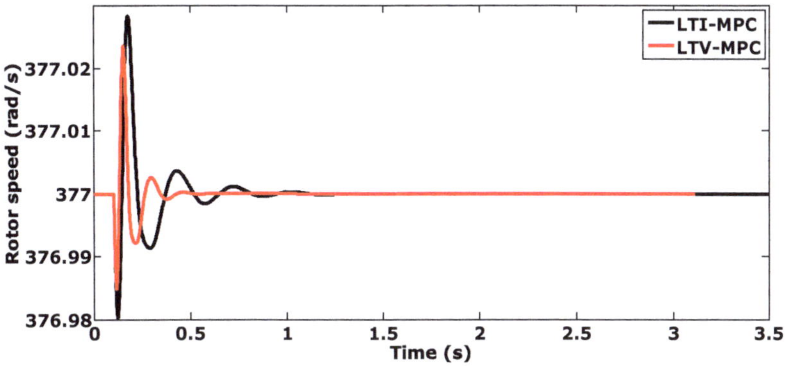

(a) Rotor speed during the PV-DG source intermittency

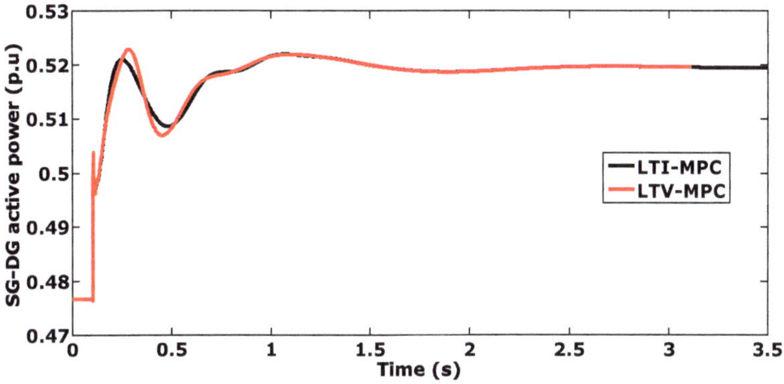

(b) SG-DG active power during the PV-DG source intermittency

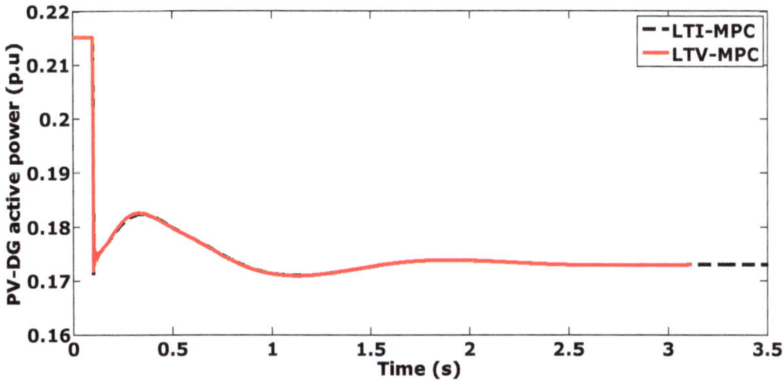

(c) PV-DG active power during the PV-DG source intermittency

Fig. 6.4 LTV-MPC performance for PV-DG source intermittency: **a** rotor speed, **b** SG-DG active power, **c** PV-DG active power, **d** SG-DG reactive power, **e** PV-DG reactive power, **f** bus B1 voltage, and **g** bus B3 voltage

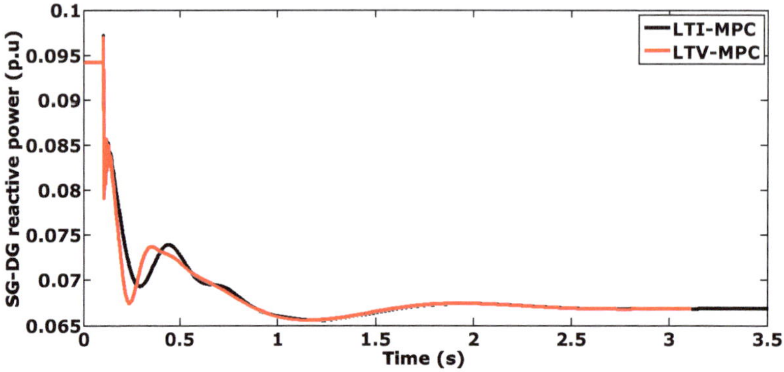

(d) SG-DG reactive power during the PV-DG source intermittency

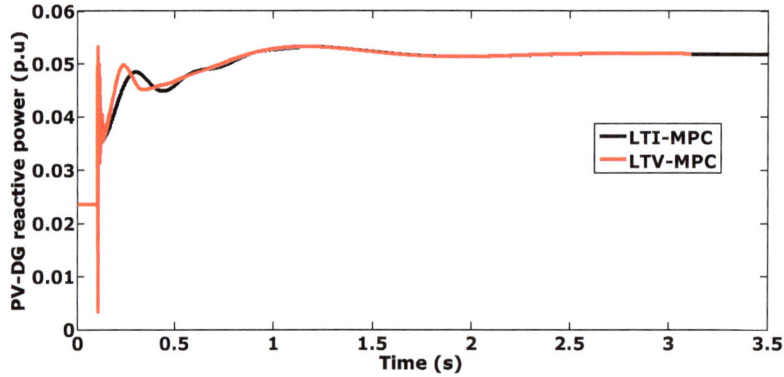

(e) PV-DG reactive power during the PV-DG source intermittency

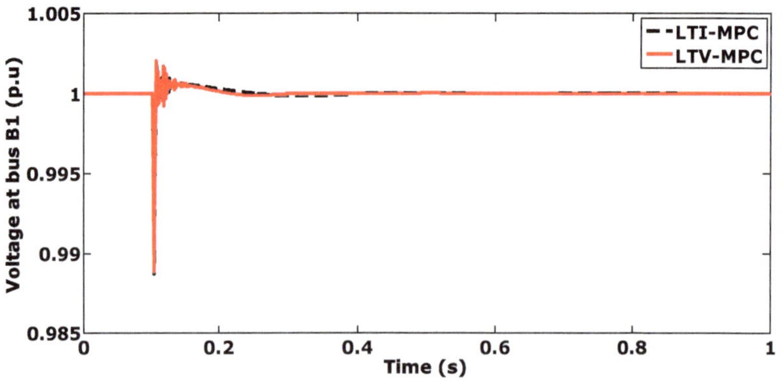

(f) Bus B1 voltage during the PV-DG source intermittency

Fig. 6.4 (continued)

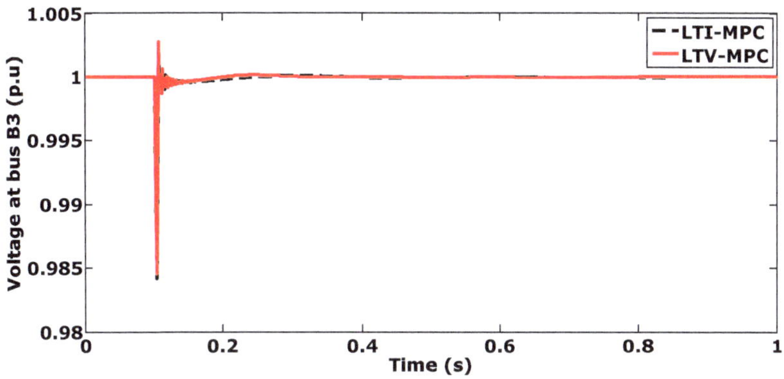

(g) Bus B3 voltage during the PV-DG source intermittency

Fig. 6.4 (continued)

References

Kouvaritakis B, Cannon M, Rossiter JA (1999) Non-linear model based predictive control. Int J Control 72(10):919–928

Puvvula V, Verma PP, Kesanakurthy SS (2020) Two-parameter Kautz network-based LTV-MPC for non-linear standalone micro-grid control. IET Renew Power Gener 14(12):2221–2231

Chapter 7
Special Functions in the MPC Formulation

Abstract In the LTI-MPC and LTV-MPC formulations discussed in the previous chapters, the number of optimal variables that are to be evaluated at each sampling instant increases with an increase in the length of the control horizon N_c and the number of control inputs n_{ip} in the micro-grid model. Orthonormal special functions are employed for approximating the original pulse operator-based control trajectories within the control horizon. The approximation aims to decrease the number of decision variables (optimal variables) in the optimal control problem without compromising the controller's performance. Two kinds of special functions, namely Laguerre functions and two-parameter Kautz functions, are employed in this chapter.

Keywords Kautz functions · Laguerre functions · Orthonormal basis · Special functions

7.1 Role of Orthonormal Special Functions in the MPC

In the formulation of the optimal control problem of the LTI-MPC and LTV-MPC, $\boldsymbol{U}$ is the control vector that is to be optimized at each sampling instant. Any input vector $\boldsymbol{\Delta u}(k_i + m)$ in $\boldsymbol{U}$ at a distance of 'm' samples from the present sampling instant k_i can be expressed using the pulse operators as follows:

$$\Delta \boldsymbol{u}(k_i + m) = \left[\boldsymbol{\delta}(m)^{\mathrm{T}} \quad \boldsymbol{\delta}(m-1)^{\mathrm{T}} \quad \boldsymbol{\delta}(m-2)^{\mathrm{T}} \ldots \boldsymbol{\delta}(m - N_c + 1)^{\mathrm{T}}\right]\boldsymbol{U} \tag{7.1}$$

$$\boldsymbol{\delta} = \left[\delta_1 \quad \delta_2 \quad \delta_3 \ldots \delta_{n_{ip}}\right]^{\mathrm{T}}, \; \delta_s(m-i) = \begin{cases} 1 \text{ for } i = m \\ 0 \text{ for } i \neq m \end{cases}, \quad \forall s \in \left[1, n_{ip}\right]$$

From (7.1), we can observe that n_{ip} pulses are required to capture $\boldsymbol{\Delta u}$ at a particular sampling instant. If the control horizon length is N_c, then the total number of pulses required to capture the entire vector $\boldsymbol{U}$ is $n_{ip}N_c$. This indicates that at each sampling instant, the optimal control problem should be evaluated online to generate $n_{ip}N_c$ optimal variables (decision variables).

P. Vidyasagar and K. Shanti Swarup, *Design and Development of Model Predictive Primary Control of Micro Grids*, Springer Tracts in Electrical and Electronics Engineering, https://doi.org/10.1007/978-981-19-5852-6_7

For large control horizon lengths, a large number of control inputs, and high sampling rates, the online evaluation of the optimal control problem of the MPC with pulse operators is time-consuming. To deal with such scenarios at the primary control level of the micro-grid, special functions are employed in this chapter (Wang 2009; Misra et al. 2016). These special functions are orthonormal basis functions and are extensively used in recent years for system identification, modelling, filtering, approximation, and control (Rosa et al. 2009; Heuberger et al. 2005; Wang 2012). They involve free parameters and forms a complete orthonormal set of functions in $l^2[0, \infty)$. Due to their assumed completeness, any arbitrary finite energy signal $f(k)$ can be approximated to any degree of accuracy by a linear combination of a finite number (N) of special functions.

$$f(k) = c_1 o_1(k) + c_2 o_2(k) + c_3 o_3(k) + \cdots + c_N o_N(k) \tag{7.2}$$

where, $\{o_1, o_2, o_3 \ldots o_N\}$ are a set of N number of special functions and $\{c_1, c_2, c_3 \ldots c_N\}$ are the coefficients to be determined in an optimal manner (decision variables). The orthonormal nature of the special functions can be expressed by the following equation:

$$\sum_{m=0}^{\infty} o_i(m) o_j(m) = \begin{cases} 1 & \text{for } i = j \\ 0 & \text{for } i \neq j \end{cases} \tag{7.3}$$

7.2 Approximation of the Original Control Trajectories

7.2.1 Laguerre Functions

Laguerre functions are one set of orthonormal basis functions with a real pole in their structure as a free parameter. They are used to approximate the control trajectories with good damping. They are very popular in system identification, modelling, and control due to their simplicity in the choice of free parameters, structure, and generation. In a Laguerre network of N functions, the z-transform of any i-th Laguerre function is given by:

$$l_i(z) = \frac{\sqrt{1-p^2}}{1-pz^{-1}} \left(\frac{z^{-1}-p}{1-pz^{-1}} \right)^{i-1} \tag{7.4}$$

where p is the real pole of the Laguerre network with $0 \leq p < 1$. From (7.4), the Laguerre functions can be generated by the following discrete state-space model (Wang 2009; Vidyasagar and Swamp 2016; Sai Sesha and Kesanakurthy 2018):

$$\boldsymbol{Lag}(k+1) = \boldsymbol{A}_l \boldsymbol{Lag}(k) \tag{7.5}$$

$$\boldsymbol{Lag}(k) = [l_1(k) \quad l_2(k) \quad l_3(k) \ldots l_N(k)]^{\mathrm{T}},$$

$$\boldsymbol{A}_l = \begin{bmatrix} p & 0 & 0 & \cdots & 0 \\ \gamma & p & 0 & \cdots & 0 \\ -p\gamma & \gamma & p & \cdots & 0 \\ \ldots & \ldots & \ldots & \ldots & \ldots \\ \ldots & \ldots & \ldots & \ldots & \ldots \\ (-p)^{N-2}\gamma & (-p)^{N-3}\gamma & \cdots & \gamma & p \end{bmatrix},$$

$$\gamma = 1 - p^2, \quad \boldsymbol{Lag}(0)^{\mathrm{T}} = \sqrt{\gamma}\left[1 \quad -p \quad p^2 \quad -p^3 \ldots (-1)^{N-1} p^{N-1}\right]$$

Any s-th input in the vector $\Delta\boldsymbol{u}(k_i + m)$ can be written as a linear combination of N_s Laguerre functions, which is given by:

$$\Delta u_s(k_i + m) = \boldsymbol{Lag}_s(m)^{\mathrm{T}} \boldsymbol{\eta}_s \tag{7.6}$$

$$\boldsymbol{Lag}_s(m)^{\mathrm{T}} = \left[l_{1s}(m) \quad l_{2s}(m) \quad l_{3s}(m) \ldots l_{N_s s}(m)\right],$$

$$\boldsymbol{\eta}_s^{\mathrm{T}} = \left[c_{1s} \quad c_{2s} \quad c_{3s} \ldots c_{N_s s}\right]$$

Similarly, we can write the entire vector $\Delta\boldsymbol{u}(k_i + m)$ as:

$$\Delta\boldsymbol{u}(k_i + m) = \boldsymbol{L}(m)^{\mathrm{T}} \boldsymbol{\eta} \tag{7.7}$$

$$\boldsymbol{\eta}^{\mathrm{T}} = \left[\boldsymbol{\eta}_1^{\mathrm{T}} \quad \boldsymbol{\eta}_2^{\mathrm{T}} \quad \boldsymbol{\eta}_3^{\mathrm{T}} \ldots \boldsymbol{\eta}_{n_{ip}}^{\mathrm{T}}\right],$$

$$\boldsymbol{L}(m)^{\mathrm{T}} = \begin{bmatrix} \boldsymbol{Lag}_1(m)^{\mathrm{T}} & 0 & 0 & 0 & \ldots & 0 \\ 0 & \boldsymbol{Lag}_2(m)^{\mathrm{T}} & 0 & 0 & \ldots & 0 \\ 0 & 0 & \boldsymbol{Lag}_3(m)^{\mathrm{T}} & 0 & \ldots & 0 \\ \ldots & \ldots & \ldots & \ldots & \ldots & \ldots \\ 0 & 0 & 0 & \ldots & \ldots & \boldsymbol{Lag}_{n_{ip}}(m)^{\mathrm{T}} \end{bmatrix}$$

7.2.2 Kautz Functions

Laguerre functions have been so popular in system identification and control due to their simplicity. The scaling factor p predominately determines the properties of the Laguerre functions. The complexity of the functions increases as the order of the network N increases. However, this simplicity can also lead to limitations in the application of Laguerre networks. Particularly when the underlying control trajectories have complex poles, the Laguerre networks require many functions to approximate such control trajectories (Wang 2009). This leads to the introduction of the Kautz networks, which can approximate finite energy signals with strong oscillatory behaviour with a reasonable number of functions (Vidyasagar et al. 2020; Puvvula et al. 2020). Kautz network allows non-identical poles in its structure. Particularly, they allow complex poles in their structure. This leads to an approximation of the oscillatory trajectories with less number of functions compared to the Laguerre networks. The two-parameter Kautz network involves a pair of complex conjugate poles p_1 and $\overline{p_1}$. A Kautz network with N Kautz functions constitutes a second-order generalization (Vidyasagar et al. 2020; Puvvula et al. 2020; Kautz 1954; Wahlberg 1994; Khan et al. 2011; Cheng et al. 2018; Rosa et al. 2007; Tanguy et al. 2002):

$$K_{2n}(z) = \frac{z\sqrt{(1-c^2)(1-b^2)}}{z^2 + b(c-1)z - c}\left(\frac{-cz^2 + b(c-1)z + 1}{z^2 + b(c-1)z - c}\right)^{n-1}, \quad n = 1, 2, 3 \ldots N/2 \tag{7.8}$$

$$K_{2n-1}(z) = \frac{z(z-b)\sqrt{(1-c^2)}}{z^2 + b(c-1)z - c}\left(\frac{-cz^2 + b(c-1)z + 1}{z^2 + b(c-1)z - c}\right)^{n-1}, \quad n = 1, 2, 3 \ldots N/2 \tag{7.9}$$

$$b = \frac{p_1 + \overline{p_1}}{1 + \overline{p_1}p_1}, \quad c = -\overline{p_1}p_1$$

The odd Kautz functions can be generated by the following discrete state-space model:

$$\boldsymbol{K_o}(k+1) = \boldsymbol{A_o}\boldsymbol{K_o}(k)$$

$$\boldsymbol{Kat_o}(k) = \boldsymbol{C_o}\boldsymbol{K_o}(k) \tag{7.10}$$

$$\boldsymbol{A_o} = \begin{bmatrix} \boldsymbol{M}_1 & 0 & 0 & \cdots & \cdots\cdots\cdots\cdots & 0 \\ \boldsymbol{M}_2 & \boldsymbol{M}_1 & 0 & 0 & \cdots\cdots\cdots\cdots & 0 \\ 0 & \boldsymbol{M}_2 & \boldsymbol{M}_1 & 0 & 0 \;\cdots\cdots\cdots & 0 \\ 0 & 0 & \boldsymbol{M}_2 & \boldsymbol{M}_1 & \cdots\cdots\cdots\cdots & 0 \\ \cdots & \cdots & \cdots & \cdots & \cdots\cdots\cdots\cdots & \cdots \\ \cdots & \cdots & \cdots & \cdots & \cdots\cdots\cdots\cdots & \cdots \\ 0 & 0 & 0 & \cdots & \cdots\cdots\cdots\; \boldsymbol{M}_2 & \boldsymbol{M}_1 \end{bmatrix} + \begin{bmatrix} 0 & 0 & 0 \;\cdots & \cdots 0 \\ 0 & 0 & 0 \;\cdots & \cdots 0 \\ -c\boldsymbol{M}_2 & 0 & 0 \;\; 0 & \cdots 0 \\ c^2\boldsymbol{M}_2 & -c\boldsymbol{M}_2 & 0 \;\; 0 & \cdots 0 \\ \cdots & \cdots & \cdots\cdots & \cdots\cdots \\ \cdots & \cdots & \cdots\cdots & \cdots\cdots \\ (-c)^{\frac{N}{2}-2}\boldsymbol{M}_2 & (-c)^{\frac{N}{2}-3}\boldsymbol{M}_2 & \cdots & -c\boldsymbol{M}_2 \; 0 \;\; 0 \end{bmatrix},$$

$$\boldsymbol{K_o} = \begin{bmatrix} K_1 & \tau_1 & K_3 & \tau_3 \ldots K_{N-1} & \tau_{N-1} \end{bmatrix}^{\mathrm{T}}, \ \boldsymbol{Kat_o} = \begin{bmatrix} K_1 & K_3 \ldots K_{N-1} \end{bmatrix}^{\mathrm{T}},$$

$$\boldsymbol{C_o} = \begin{bmatrix} 1\,0\,1\,0 \ldots 1\,0 \end{bmatrix}^{\mathrm{T}}, \boldsymbol{M}_1 = \begin{bmatrix} 0 & 1 \\ c & -b(c-1) \end{bmatrix},$$

$$\boldsymbol{M}_2 = \begin{bmatrix} 0 & 0 \\ 1-c^2 & bc(c-1)+b(c-1) \end{bmatrix}$$

Similarly, even Kautz functions can be generated by the following discrete state-space model:

$$\boldsymbol{K_e}(k+1) = \boldsymbol{A_e K_e}(k)$$

$$\boldsymbol{Kat_e}(k) = \boldsymbol{C_e K_e}(k) \tag{7.11}$$

$$\boldsymbol{A_e} = \boldsymbol{A_o}, \ \boldsymbol{K_e} = \begin{bmatrix} K_2 & \tau_2 & K_4 & \tau_4 \ldots K_N & \tau_N \end{bmatrix}^{\mathrm{T}}, \ \boldsymbol{Kat_e} = \begin{bmatrix} K_2 & K_4 \ldots K_N \end{bmatrix}^{\mathrm{T}}, \ \boldsymbol{C_e} = \boldsymbol{C_o}$$

The initial values $\boldsymbol{K_o}(0)$ and $\boldsymbol{K_e}(0)$ can be found by solving the corresponding difference equations for (7.8) and (7.9). The entire Kautz network with N Kautz functions can be described by:

$$\boldsymbol{Kat} = \begin{bmatrix} K_1 & K_2 & K_3 & K_4 \ldots K_{N-1} & K_N \end{bmatrix}^{\mathrm{T}} \tag{7.12}$$

The objective here is to approximate any s-th input in the vector $\Delta\boldsymbol{u}(k_i+m)$ by a network of N_s Kautz functions, which is given by:

$$\Delta u_s(k_i+m) = \boldsymbol{Kat_s}(m)^{\mathrm{T}} \boldsymbol{\eta_s} \tag{7.13}$$

$$\boldsymbol{Kat_s}(m)^{\mathrm{T}} = \begin{bmatrix} K_{1s}(m) & K_{2s}(m) & K_{3s}(m) \ldots K_{N_s s}(m) \end{bmatrix}, \ \boldsymbol{\eta_s^{\mathrm{T}}} = \begin{bmatrix} c_{1s} & c_{2s} & c_{3s} \ldots c_{N_s s} \end{bmatrix}$$

Similarly, we can write the entire vector $\Delta\boldsymbol{u}(k_i+m)$ as:

$$\Delta\boldsymbol{u}(k_i+m) - \boldsymbol{K}(m)^{\mathrm{T}} \boldsymbol{\eta} \tag{7.14}$$

$$\boldsymbol{\eta}^{\mathrm{T}} = \begin{bmatrix} \boldsymbol{\eta}_1^{\mathrm{T}} & \boldsymbol{\eta}_2^{\mathrm{T}} & \boldsymbol{\eta}_3^{\mathrm{T}} \ldots \boldsymbol{\eta}_{\boldsymbol{n_{ip}}}^{\mathrm{T}} \end{bmatrix},$$

$$\boldsymbol{K}(m)^{\mathrm{T}} = \begin{bmatrix} \boldsymbol{Kat}_1(m)^{\mathrm{T}} & 0 & 0 & 0 & \ldots & 0 \\ 0 & \boldsymbol{Kat}_2(m)^{\mathrm{T}} & 0 & 0 & \ldots & 0 \\ 0 & 0 & \boldsymbol{Kat}_3(m)^{\mathrm{T}} & 0 & \ldots & 0 \\ \ldots & \ldots & \ldots & \ldots & \ldots & \ldots \\ 0 & 0 & 0 & & \ldots & \boldsymbol{Kat}_{\boldsymbol{n_{ip}}}(m)^{\mathrm{T}} \end{bmatrix}$$

7.3 Mathematical Formulation of the LTI-MPC Using Special Functions

7.3.1 Augmented Model

The augmented model of the micro-grid for the formulation of the LTI-MPC is already discussed in Chap. 5. For the understanding purpose, it is briefly presented below. The nonlinear mathematical model of the micro-grid is given by:

$$
\begin{aligned}
\frac{\mathrm{d}\boldsymbol{X}(t)}{\mathrm{d}t} &= \boldsymbol{f}(\boldsymbol{X}(t), \boldsymbol{V}(t)) \\
\boldsymbol{Y}(t) &= \boldsymbol{g}(\boldsymbol{X}(t))
\end{aligned} \tag{7.15}
$$

The linear time-invariant (LTI) approximation of the nonlinear micro-grid model (7.15) around the present operating point $(\boldsymbol{X}(t_s), \boldsymbol{V}(t_s - T_s))$ is given by:

$$
\begin{aligned}
\frac{\mathrm{d}(\Delta\boldsymbol{X}(t))}{\mathrm{d}t} &= \boldsymbol{f}(\boldsymbol{X}(t_s), \boldsymbol{V}(t_s - T_s)) + \frac{\partial \boldsymbol{f}}{\partial \boldsymbol{X}}\Big|_{\substack{X = X(t_s) \\ V = V(t_s - T_s)}} \Delta\boldsymbol{X}(t) \\
&\quad + \frac{\partial \boldsymbol{f}}{\partial \boldsymbol{V}}\Big|_{\substack{X = X(t_s) \\ V = V(t_s - T_s)}} \Delta\boldsymbol{V}(t) \\
\boldsymbol{Y}(t) &= \boldsymbol{g}(\boldsymbol{X}(t_s)) + \frac{\partial \boldsymbol{g}}{\partial \boldsymbol{X}}\Big|_{\substack{X = X(t_s) \\ V = V(t_s - T_s)}} \Delta\boldsymbol{X}(t) \\
\Delta\boldsymbol{V}(t) &= \boldsymbol{V}(t) - \boldsymbol{V}(t_s - T_s), \quad \Delta\boldsymbol{X}(t) = \boldsymbol{X}(t) - \boldsymbol{X}(t_s)
\end{aligned} \tag{7.16}
$$

The linear model (7.16) can be written as:

$$
\begin{aligned}
\frac{\mathrm{d}\boldsymbol{Z}_c(t)}{\mathrm{d}t} &= \overline{\boldsymbol{Z}} + \boldsymbol{A}_t \boldsymbol{Z}_c(t) + \boldsymbol{B}_t \boldsymbol{u}(t) \\
\boldsymbol{Y}(t) &= \overline{\boldsymbol{Y}} + \boldsymbol{C}_t \boldsymbol{Z}_c(t) \\
\boldsymbol{Z}_c(t) &= \Delta\boldsymbol{X}(t),\ \overline{\boldsymbol{Z}} = \boldsymbol{f}(\boldsymbol{X}(t_s), \boldsymbol{V}(t_s - T_s)),\ \overline{\boldsymbol{Y}} = \boldsymbol{g}(\boldsymbol{X}(t_s)),\ \boldsymbol{u}(t) = \Delta\boldsymbol{V}(t) \\
\boldsymbol{A}_t &= \frac{\partial \boldsymbol{f}}{\partial \boldsymbol{X}}\Big|_{\substack{X = X(t_s) \\ V = V(t_s - T_s)}},\quad \boldsymbol{B}_t = \frac{\partial \boldsymbol{f}}{\partial \boldsymbol{V}}\Big|_{\substack{X = X(t_s) \\ V = V(t_s - T_s)}}, \\
\boldsymbol{C}_t &= \frac{\partial \boldsymbol{g}}{\partial \boldsymbol{X}}\Big|_{\substack{X = X(t_s) \\ V = V(t_s - T_s)}}
\end{aligned} \tag{7.17}
$$

The continuous model (7.17) is then converted to a discrete model using zero-order hold (ZOH):

$$\begin{aligned} \boldsymbol{Z}_c(k+1) &= \tilde{\boldsymbol{Z}} + \boldsymbol{A}_d \boldsymbol{Z}_c(k) + \boldsymbol{B}_d \boldsymbol{u}(k) \\ \boldsymbol{Y}(k) &= \tilde{\boldsymbol{Y}} + \boldsymbol{C}_d \boldsymbol{Z}_c(k) \end{aligned} \tag{7.18}$$

$(\boldsymbol{A}_d, \boldsymbol{B}_d, \boldsymbol{C}_d)$ is the discrete-time state-space triplet with $\{k = 0 \; implies \; that \; t = t_s\}$. $\tilde{\boldsymbol{Z}}, \tilde{\boldsymbol{Y}}$ are discrete counterparts to $\overline{\boldsymbol{Z}}, \overline{\boldsymbol{Y}}$. As the micro-grid control requires an integral action, the integrators are embedded into (7.18), converting it to an augmented model given by:

$$\begin{aligned} \boldsymbol{Z}(k+1) &= \overline{\boldsymbol{\theta}} + \boldsymbol{A}\boldsymbol{Z}(k) + \boldsymbol{B}\Delta\boldsymbol{u}(k) \\ \boldsymbol{Y}(k) &= \tilde{\boldsymbol{Y}} + \boldsymbol{C}\boldsymbol{Z}(k) \\ \overline{\boldsymbol{\theta}}^{\mathrm{T}} &= \begin{bmatrix} \tilde{\boldsymbol{Z}}^{\mathrm{T}} & 0_{1\times n_{ip}} \end{bmatrix}, \; \boldsymbol{Z}(k)^{\mathrm{T}} = \begin{bmatrix} \boldsymbol{Z}_c(k)^{\mathrm{T}} & \boldsymbol{u}(k-1)^{\mathrm{T}} \end{bmatrix}, \\ \boldsymbol{A} &= \begin{bmatrix} \boldsymbol{A}_d & \boldsymbol{B}_d \\ 0_{n_{ip}\times n} & \boldsymbol{I}_{n_{ip}\times n_{ip}} \end{bmatrix}, \; \boldsymbol{B} = \begin{bmatrix} \boldsymbol{B}_d \\ \boldsymbol{I}_{n_{ip}\times n_{ip}} \end{bmatrix}; \\ \boldsymbol{C} &= \begin{bmatrix} \boldsymbol{C}_d & 0_{n_{op}\times n_{ip}} \end{bmatrix}, \; \Delta\boldsymbol{u}(k) = \boldsymbol{u}(k) - \boldsymbol{u}(k-1) \end{aligned} \tag{7.19}$$

7.3.2 LTI-MPC Using Special Functions

Let $\boldsymbol{U}$ be a vector that contains the future control trajectories from the present sampling instant $k = k_i$ to a control horizon of length N_c samples. Let $\boldsymbol{Y_E}$ be a vector with micro-grid outputs predicted from the sampling instant $k = k_i + 1$ to a prediction horizon of length N_p samples.

$$\boldsymbol{Y_E} = \begin{bmatrix} \boldsymbol{Y}(k_i+1)^{\mathrm{T}} & \boldsymbol{Y}(k_i+2)^{\mathrm{T}} & \ldots \boldsymbol{Y}(k_i+N_P)^{\mathrm{T}} \end{bmatrix}^{\mathrm{T}} \tag{7.20}$$

$$\boldsymbol{U} = \begin{bmatrix} \Delta\boldsymbol{u}(k_i)^{\mathrm{T}} & \Delta\boldsymbol{u}(k_i+1)^{\mathrm{T}} \ldots \Delta\boldsymbol{u}(k_i+N_C-1)^{\mathrm{T}} \end{bmatrix}^{\mathrm{T}} \tag{7.21}$$

From (7.19), we can write the following equation:

$$\boldsymbol{Y}(k_i+m) = \tilde{\boldsymbol{Y}} + \boldsymbol{C}\boldsymbol{A}^m\boldsymbol{Z}(k_i) + \boldsymbol{C}\sum_{j=0}^{m-1}\boldsymbol{A}^j\overline{\boldsymbol{\theta}} + \boldsymbol{C}\sum_{j=0}^{m-1}\boldsymbol{A}^{m-1-j}\boldsymbol{B}\Delta\boldsymbol{u}(k_i+j) \tag{7.22}$$

Within the prediction horizon, the cost function associated with the LTI-MPC is quadratic, which is given by:

$$J = \sum_{m=1}^{N_p} (\boldsymbol{w}(k_i) - \boldsymbol{Y}(k_i + m))^{\mathrm{T}} (\boldsymbol{w}(k_i) - \boldsymbol{Y}(k_i + m)) + \sum_{m=0}^{N_c - 1} \Delta \boldsymbol{u}(k_i + m)^{\mathrm{T}} \boldsymbol{r}(k_i) \Delta \boldsymbol{u}(k_i + m) \tag{7.23}$$

$\boldsymbol{w}(k_i)$ consists of set-points for each output in the output vector $\boldsymbol{Y}(k_i + m)$. $\boldsymbol{r}(k_i)$ consists of penalties on each input increment in the vector $\Delta\boldsymbol{u}(k_i + m)$. From (7.7) and (7.14), we can write the input vector $\Delta\boldsymbol{u}(k_i + m)$ as:

$$\Delta \boldsymbol{u}(k_i + m) = \boldsymbol{O}(m)^{\mathrm{T}} \boldsymbol{\eta}, \quad \boldsymbol{O} = \boldsymbol{L}(\boldsymbol{or})\boldsymbol{K} \tag{7.24}$$

Using (7.24), we can write the following equation:

$$\sum_{m=0}^{N_c - 1} \Delta \boldsymbol{u}(k_i + m)^{\mathrm{T}} \boldsymbol{r}(k_i) \Delta \boldsymbol{u}(k_i + m) = \sum_{m=0}^{N_c - 1} \boldsymbol{\eta}^{\mathrm{T}} \boldsymbol{O}(m) \boldsymbol{r}(k_i) \boldsymbol{O}(m)^{\mathrm{T}} \boldsymbol{\eta} \tag{7.25}$$

Since the special functions are orthonormal, we have:

$$\sum_{m=0}^{\infty} o_i(m) o_j(m) = \begin{Bmatrix} 1, \text{ if } i = j \\ 0, \text{ if } i \neq j \end{Bmatrix} \tag{7.26}$$

Applying orthonormal property (7.26) on (7.25) for sufficiently large control horizon N_c, we can write the following:

$$\sum_{m=0}^{N_c - 1} \Delta \boldsymbol{u}(k_i + m)^{\mathrm{T}} \boldsymbol{r}(k_i) \Delta \boldsymbol{u}(k_i + m) \cong \boldsymbol{\eta}^{\mathrm{T}} \boldsymbol{R}_{\boldsymbol{w}} \boldsymbol{\eta} \tag{7.27}$$

$\boldsymbol{R}_w$ is the penalty matrix on input increments. Substituting (7.22), (7.24), and (7.27) in the quadratic cost function (7.23), we get the following convex quadratic optimal control problem:

$$\begin{aligned}
&\text{Min} J = \boldsymbol{\eta}^{\mathrm{T}} \boldsymbol{E} \boldsymbol{\eta} + 2\boldsymbol{\eta}^{\mathrm{T}} (-\boldsymbol{E}_1 + \boldsymbol{E}_2 + \boldsymbol{E}_3 + \boldsymbol{E}_4) + \text{non } \boldsymbol{\eta} \text{ terms} \\
&\text{S. T.} \quad \Delta \boldsymbol{u}_{\min} \le \Delta \boldsymbol{u}(k_i) \le \Delta \boldsymbol{u}_{\max}, \boldsymbol{u}_{\min} \le \boldsymbol{u}(k_i) \le \boldsymbol{u}_{\max}, \boldsymbol{V}_{\min} \le \boldsymbol{V}(k_i) \le \boldsymbol{V}_{\max}, \\
&P_{G1,\min} \le P_{G1}(k_i + 1) \le P_{G1,\max}, Q_{G1,\min} \le Q_{G1}(k_i + 1) \le Q_{G1,\max}, \\
&P_{G3,\min} \le P_{G3}(k_i + 1) \le P_{G3,\max}, Q_{G3,\min} \le Q_{G3}(k_i + 1) \le Q_{G3,\max}, \\
&V_{\text{mppt}}(G, T) \le V_{\text{dc}}(k_i + 1) \le V_{oc}(G, T), \boldsymbol{E} = \boldsymbol{R}_{\boldsymbol{w}} + \sum_{m=1}^{N_p} \phi(m) \boldsymbol{Q} \phi(m)^{\mathrm{T}}, \ \boldsymbol{Q} = \boldsymbol{C}^{\mathrm{T}} \boldsymbol{C}, \\
&\boldsymbol{E}_1 = \sum_{m=1}^{N_p} \phi(m) \boldsymbol{C}^{\mathrm{T}} \boldsymbol{w}(k_i), \ \boldsymbol{E}_2 = \sum_{m=1}^{N_p} \phi(m) \boldsymbol{C}^{\mathrm{T}} \tilde{\boldsymbol{Y}}, \ \boldsymbol{E}_3 = \sum_{m=1}^{N_p} \phi(m) \boldsymbol{Q} \boldsymbol{F}(m) \overline{\boldsymbol{\theta}},
\end{aligned}$$

$$\boldsymbol{F}(m)=\sum_{j=0}^{m-1}\boldsymbol{A}^{j},\ \boldsymbol{E}_4=\sum_{m=1}^{N_p}\phi(m)\boldsymbol{Q}\boldsymbol{A}^{m}\boldsymbol{Z}(k_i),\ \phi(m)^{\mathrm{T}}=\sum_{j=0}^{m-1}\boldsymbol{A}^{m-1-j}\boldsymbol{B}\boldsymbol{O}(j)^{\mathrm{T}} \tag{7.28}$$

$(P_{G1,\min}, P_{G1,\max})$, $(Q_{G1,\min}, Q_{G1,\max})$, $(P_{G3,\min}, P_{G3,\max})$, $(Q_{G3,\min}, Q_{G3,\max})$ are minimum and maximum limits of the active and reactive powers of the SG-DG and PV-DG. $V_{\mathrm{mppt}}(G, T)$ and $V_{oc}(G, T)$ are the output voltages of the PV array corresponding to the maximum power point and open circuit. The convex quadratic optimal control problem (7.28) is evaluated, and an optimal coefficient vector $\boldsymbol{\eta}=\boldsymbol{\eta}_{\mathbf{opt}}$ is generated. By applying the receding horizon principle, inputs corresponding to the present sampling instant $\Delta\boldsymbol{u}(k_i)=\boldsymbol{O}(0)^{T}\boldsymbol{\eta}_{\mathbf{opt}}$ are calculated. When the next sample arrives, the whole procedure described above is repeated. The flowchart of the operation of the LTI-MPC with special functions is shown in Fig. 7.1.

7.4 Mathematical Formulation of the LTV-MPC Using Special Functions

7.4.1 Augmented Model

The prediction horizon in the continuous-time domain is $\left(t_s\le t\le t_s+N_pT_s\right)$, and in the discrete-time domain is given by $\left(k_i\le k\le k_i+N_p\right)$. The control horizon is given by $(t_s\le t\le t_s+(N_c-1)T_s)$ in continuous-time, and the discrete-time control horizon is given by $(k_i\le k\le k_i+N_c-1)$. Within the prediction horizon, let us consider the following:

$$\boldsymbol{X}(t)=\boldsymbol{X}_{\mathbf{ref}}(t)+\Delta\boldsymbol{X}(t),\quad \left(t_s\le t\le t_s+N_pT_s\right) \tag{7.29}$$

$$\boldsymbol{V}(t)=\boldsymbol{V}_{\mathbf{ref}}(t)+\Delta\boldsymbol{V}(t),\quad \left(t_s\le t\le t_s+N_pT_s\right) \tag{7.30}$$

$$\boldsymbol{Y}(t)=\boldsymbol{Y}_{\mathbf{ref}}(t)+\Delta\boldsymbol{Y}(t),\quad \left(t_s\le t\le t_s+N_pT_s\right) \tag{7.31}$$

$\boldsymbol{X}_{\mathbf{ref}}, \boldsymbol{V}_{\mathbf{ref}}, \boldsymbol{Y}_{\mathbf{ref}}$ are the reference trajectories of the state vector, input vector, and output vector of the micro-grid. $\Delta\boldsymbol{X}$, $\Delta\boldsymbol{V}$, $\Delta\boldsymbol{Y}$ are the perturbations around their respective reference trajectories. The linear time-variant (LTV) approximation of the micro-grid model (7.15) around the reference trajectories (7.29), (7.30), and (7.31) is given by:

$$\frac{\mathrm{d}\boldsymbol{Z}_c(t)}{\mathrm{d}t}=\boldsymbol{A}_c(t)\boldsymbol{Z}_c(t)+\boldsymbol{B}_c(t)\boldsymbol{u}(t)$$
$$\Delta\boldsymbol{Y}(t)=\boldsymbol{C}_c(t)\boldsymbol{Z}_c(t),\quad \left(t_s\le t\le t_s+N_pT_s\right)$$

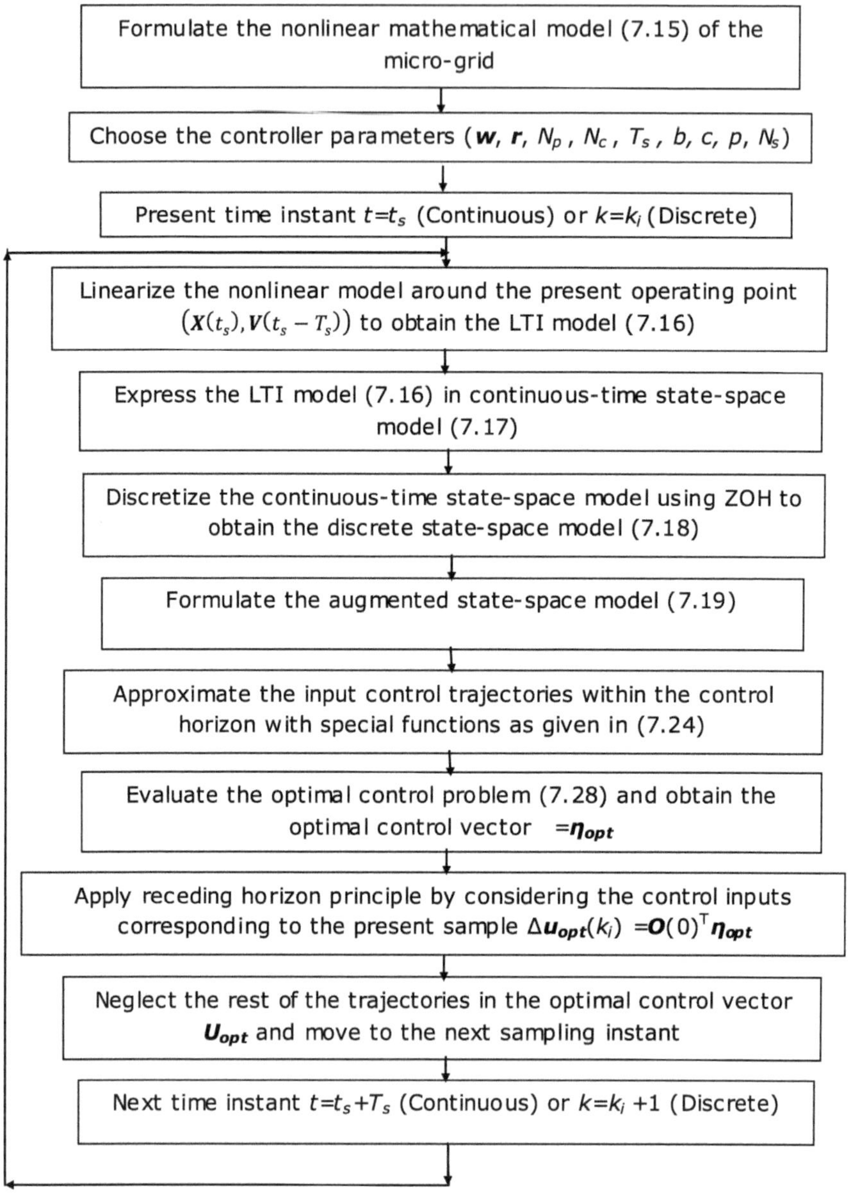

Fig. 7.1 Flowchart of the operation of the LTI-MPC with special functions

$$Z_c(t) = \Delta X(t), u(t) = \Delta V(t), A_c(t) = \left.\frac{\partial f}{\partial X}\right|_{\substack{X(t) = X_{ref}(t) \\ V(t) = V_{ref}(t)}}, B_c(t) = \left.\frac{\partial f}{\partial V}\right|_{\substack{X(t) = X_{ref}(t) \\ V(t) = V_{ref}(t)}}$$

$$C_c(t) = \left.\frac{\partial g}{\partial X}\right|_{\substack{X(t) = X_{ref}(t) \\ V(t) = V_{ref}(t)}} \tag{7.32}$$

The LTV model (7.32) can be discretized using the zero-order hold (ZOH) with a sampling time of T_s seconds, which is given by:

$$\begin{aligned} Z_c(k+1) &= A_d(k)Z_c(k) + B_d(k)u(k) \\ \Delta Y(k) &= C_d(k)Z_c(k), \quad \left(k_i \le k \le k_i + N_p\right) \end{aligned} \tag{7.33}$$

To incorporate the integral action into the LTV-MPC, the LTV model (7.33) is augmented, resulting in a new state vector $Z(k)$:

$$\begin{aligned} Z(k+1) &= A(k)Z(k) + B(k)\Delta u(k) \\ \Delta Y(k) &= C(k)Z(k), \quad \left(k_i \le k \le k_i + N_p\right) \end{aligned} \tag{7.34}$$

$$\begin{aligned} A &= \begin{bmatrix} A_d & B_d \\ 0_{n_{ip} \times n} & I_{n_{ip} \times n_{ip}} \end{bmatrix}, \; B = \begin{bmatrix} B_d \\ I_{n_{ip} \times n_{ip}} \end{bmatrix}, \\ C &= \begin{bmatrix} C_d & 0_{n_{op} \times n_{ip}} \end{bmatrix}, \; Z(k) = \begin{bmatrix} Z_c(k)^{\mathrm{T}} & u(k-1)^{\mathrm{T}} \end{bmatrix}^{\mathrm{T}} \\ \Delta u(k) &= u(k) - u(k-1) \end{aligned}$$

7.4.1.1 Prediction of the Forced Response

Let ΔY_p be the vector that contains the predicted values of the forced response of the micro-grid within the prediction horizon:

$$\Delta Y_p = \left[\Delta Y(k_i+1)^{\mathrm{T}} \quad \Delta Y(k_i+2)^{\mathrm{T}} \ldots \Delta Y\left(k_i + N_p\right)^{\mathrm{T}}\right]^{\mathrm{T}} \tag{7.35}$$

Let U be the vector that contains the future control trajectories within the control horizon:

$$U = \left[\Delta u(k_i)^{\mathrm{T}} \quad \Delta u(k_i+1)^{\mathrm{T}} \ldots \Delta u(k_i + N_c - 1)^{\mathrm{T}}\right]^{\mathrm{T}} \tag{7.36}$$

From (7.34), we can write $\Delta Y(k_i + m)$ as:

$$\Delta Y(k_i + m) = C(k_i + m)Z(k_i + m) \tag{7.37}$$

Also, from (7.34), we can write $Z(k_i + m)$ as:

$$
\begin{aligned}
\boldsymbol{Z}(k_i+m) &= \left[\prod_{j=0}^{m-1}\boldsymbol{A}(k_i+m-1-j)\right]\boldsymbol{Z}(k_i) \\
&\quad + \sum_{j=1}^{m}\boldsymbol{A}_{j,m}\boldsymbol{B}(k_i+j-1)\Delta\boldsymbol{u}(k_i+j-1) \\
\boldsymbol{A}_{j,m} &= \begin{cases}\prod_{l=1}^{m-j}\boldsymbol{A}(k_i+m-l) & \text{for } m>j \\ I & \text{for } m=j\end{cases}
\end{aligned}
\tag{7.38}
$$

$\boldsymbol{Z}(k_i)$ is the augmented state vector at the present time instant $k = k_i$, which is zero in this case. Using (7.36), (7.37), and (7.38), the predicted vector (7.35) of the forced response of the micro-grid can be expressed as:

$$
\Delta\boldsymbol{Y}_p = \boldsymbol{\varphi}_1\boldsymbol{Z}(k_i) + \boldsymbol{\varphi}_2\boldsymbol{U} \tag{7.39}
$$

7.4.2 Prediction of the Natural Response

The natural response of the micro-grid within the prediction horizon can be calculated by solving the nonlinear model (7.15) along the state and input reference trajectories.

$$
\begin{aligned}
\boldsymbol{X}_{\mathbf{ref}}(t) &= \boldsymbol{X}(t_s) + \int_{t_s}^{t}\boldsymbol{f}(\boldsymbol{X}(t),\boldsymbol{V}_{\mathbf{ref}}(t))\mathrm{d}t \\
\boldsymbol{Y}_{\mathbf{ref}}(t) &= \boldsymbol{g}(\boldsymbol{X}_{\mathbf{ref}}(t)), \quad \left(t_s \le t \le t_s + N_pT_s\right)
\end{aligned}
\tag{7.40}
$$

The continuous-time natural response trajectories of the micro-grid can be discretized using ZOH. Let Y_p be the vector that contains the predicted values of the natural response of the micro-grid within the prediction horizon:

$$
\boldsymbol{Y}_p = \left[\boldsymbol{Y}_{\mathbf{ref}}(k_i+1)^{\mathrm{T}} \quad \boldsymbol{Y}_{\mathbf{ref}}(k_i+2)^{\mathrm{T}}\ldots\boldsymbol{Y}_{\mathbf{ref}}\left(k_i+N_p\right)^{\mathrm{T}}\right]^{\mathrm{T}} \tag{7.41}
$$

7.4.3 LTV-MPC Using Special Functions

The complete predicted response of the micro-grid within the prediction horizon is given by the sum of the natural and forced response of the micro-grid:

$$\boldsymbol{Y}_E = \boldsymbol{Y}_p + \Delta \boldsymbol{Y}_p = \boldsymbol{Y}_p + \boldsymbol{\varphi}_1 \boldsymbol{Z}(k_i) + \boldsymbol{\varphi}_2 \boldsymbol{U} \tag{7.42}$$

Substituting (7.24) in (7.36), we can write the vector $\boldsymbol{U}$ as:

$$\boldsymbol{U} = [\boldsymbol{O}(0) \quad \boldsymbol{O}(1) \quad \boldsymbol{O}(2) \ldots \boldsymbol{O}(N_c - 1) \quad]^{\mathrm{T}} \boldsymbol{\eta} = \boldsymbol{\varphi}_3^{\mathrm{T}} \boldsymbol{\eta}, \quad \boldsymbol{O} = \boldsymbol{L}(\boldsymbol{or})\boldsymbol{K} \tag{7.43}$$

Substituting (7.43) in (7.42), the complete predicted response of the micro-grid can be written as:

$$\boldsymbol{Y}_E = \boldsymbol{Y}_p + \boldsymbol{\varphi}_1 \boldsymbol{Z}(k_i) + \boldsymbol{\varphi}_2 \boldsymbol{\varphi}_3^{\mathrm{T}} \boldsymbol{\eta} \tag{7.44}$$

At each sampling instant, once $\boldsymbol{Y}_E$ is formed from (7.44), a quadratic objective function 'J' is formulated using $\boldsymbol{Y}_E$ and is given by:

$$J = (\boldsymbol{W} - \boldsymbol{Y}_E)^T (\boldsymbol{W} - \boldsymbol{Y}_E) + \boldsymbol{U}^T \boldsymbol{R} \boldsymbol{U} \tag{7.45}$$

$\boldsymbol{W}$ is a vector with reference set-points for the micro-grid outputs within the prediction horizon. $\boldsymbol{R}$ is a positive definite weight matrix on the input increments. Substituting (7.43) and (7.44) in (7.45), the complete quadratic optimal control problem is given by:

$$\begin{aligned}
&\mathrm{Min} J = \boldsymbol{\eta}^{\mathrm{T}} \boldsymbol{E} \boldsymbol{\eta} + 2\boldsymbol{\eta}^{\mathrm{T}} (\boldsymbol{E}_1 + \boldsymbol{E}_2 - \boldsymbol{E}_3) + \textbf{Constant} \\
&\text{S.T. } \boldsymbol{E} = \boldsymbol{\varphi}_3 \boldsymbol{\varphi}_2^{\mathrm{T}} \boldsymbol{\varphi}_2 \boldsymbol{\varphi}_3^{\mathrm{T}} + \boldsymbol{\varphi}_3 \boldsymbol{R} \boldsymbol{\varphi}_3^{\mathrm{T}}, \boldsymbol{E}_1 = \boldsymbol{\varphi}_3 \boldsymbol{\varphi}_2^{\mathrm{T}} \boldsymbol{\varphi}_1 \boldsymbol{Z}(k_i), \boldsymbol{E}_2 = \boldsymbol{\varphi}_3 \boldsymbol{\varphi}_2^{\mathrm{T}} \boldsymbol{Y}_p, \boldsymbol{E}_3 = \boldsymbol{\varphi}_3 \boldsymbol{\varphi}_2^{\mathrm{T}} \boldsymbol{W}, \\
&\boldsymbol{u}_{\min} \le \boldsymbol{u}(k_i) \le \boldsymbol{u}_{\max}, \boldsymbol{V}_{\min} \le \boldsymbol{V}(k_i) \le \boldsymbol{V}_{\max}, \Delta \boldsymbol{u}_{\min} \le \Delta \boldsymbol{u}(k_i) \le \Delta \boldsymbol{u}_{\max}, \\
&P_{G1,\min} \le P_{G1}(k_i + 1) \le P_{G1,\max}, Q_{G1,\min} \le Q_{G1}(k_i + 1) \le Q_{G1,\max}, \\
&P_{G3,\min} \le P_{G3}(k_i + 1) \le P_{G3,\max}, Q_{G3,\min} \le Q_{G3}(k_i + 1) \le Q_{G3,\max}, \\
&V_{\mathrm{mppt}}(G, T) \le V_{\mathrm{dc}}(k_i + 1) \le V_{oc}(G, T)
\end{aligned} \tag{7.46}$$

The pairs (P_{G1}, Q_{G1}), $(P_{G1,\max}, Q_{G1,\max})$, $(P_{G1,\min}, Q_{G1,\min})$ are active and reactive power outputs of the SG-DG, their upper and lower limits. The pairs (P_{G3}, Q_{G3}), $(P_{G3,\max}, Q_{G3,\max})$, $(P_{G3,\min}, Q_{G3,\min})$ are active and reactive power outputs of the PV-DG, their upper and lower limits. $V_{\mathrm{mppt}}(G, T)$ and $V_{oc}(G, T)$ are the output voltages of the PV array corresponding to the maximum power point and open circuit. The above quadratic optimal control problem (7.46) is evaluated (minimized) to generate an optimal coefficient vector $\eta = \boldsymbol{\eta}_{\mathbf{opt}}$. The optimal input vector corresponding to the present sampling instant $\Delta \boldsymbol{u}_{\mathbf{opt}}(k_i)$ is given by $\Delta \boldsymbol{u}_{\mathbf{opt}}(k_i) = \boldsymbol{O}(0)^{\mathrm{T}} \boldsymbol{\eta}_{\mathbf{opt}}$. From $\Delta \boldsymbol{u}_{\mathbf{opt}}(k_i)$, we can calculate the $\boldsymbol{u}_{\mathbf{opt}}(k_i)$ and $\boldsymbol{V}_{\mathbf{opt}}(k_i)$. Due to the receding horizon principle, the optimal inputs corresponding to the present time $\boldsymbol{V}_{\mathbf{opt}}(k_i) = \boldsymbol{V}_{\mathbf{opt}}(t_s)$ are applied to the micro-grid neglecting rest of the optimal control trajectories. When the next sample arrives, the whole procedure described above is repeated. The flowchart of the operation of the LTV-MPC with special functions is shown in Fig. 7.2.

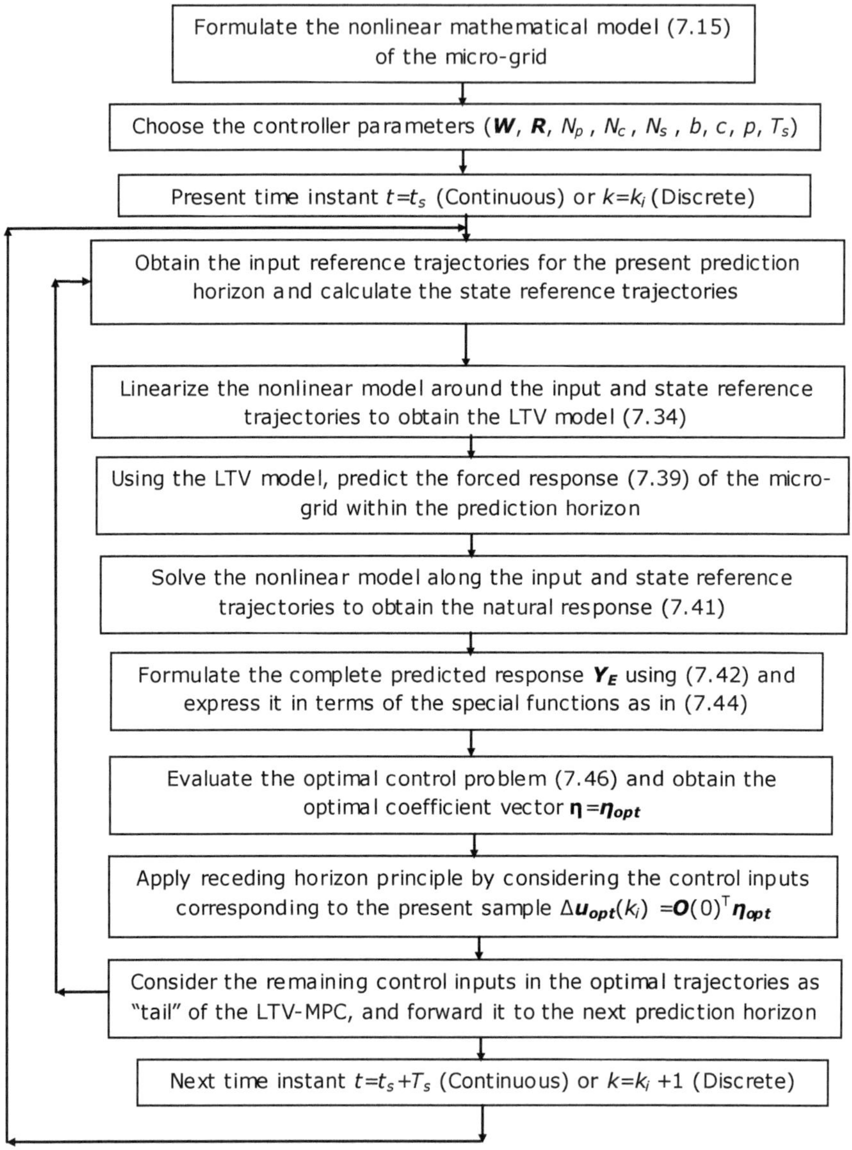

Fig. 7.2 Flowchart of the operation of the LTV-MPC with special functions

7.4.4 Choice of the Input Reference Trajectories $V_{\text{ref}}(t)$

Let us consider a vector $\boldsymbol{V}_{\text{tail}}$, which is the "tail" of the optimal control trajectories that were calculated and unused in the previous sampling instant ($k = k_i - 1$) due to the receding horizon strategy:

$$\boldsymbol{V}_{\mathbf{tail}} = [\boldsymbol{V}_{\mathbf{opt}}(k_i|k_i-1)^{\mathrm{T}}\boldsymbol{V}_{\mathbf{opt}}(k_i+1|k_i-1)^{\mathrm{T}}\ldots\boldsymbol{V}_{\mathbf{opt}}(k_i+N_c-1|k_i-1)^{\mathrm{T}}]^{\mathrm{T}} \tag{7.47}$$

The optimal inputs in the vector $\boldsymbol{V}_{\mathbf{tail}}$ are considered to be the input reference trajectories for the present control horizon ($k_i \le k \le k_i + N_c - 1$), corresponding to the present sampling instant ($k = k_i$).

$$\boldsymbol{V}_{\mathbf{ref}}(k_i+m) = \boldsymbol{V}_{\mathbf{opt}}(k_i+m|k_i-1), (0 \le m \le N_c - 1) \tag{7.48}$$

Since a zero-order hold (ZOH) circuit is used for the discretization, we can write the input reference trajectories in the continuous-time domain as follows:

$$\boldsymbol{V}_{\mathbf{ref}}(t) = \boldsymbol{V}_{\mathbf{opt}}(k_i+m|k_i-1), t_s + mT_s \le t < t_s + (m+1)T_s \tag{7.49}$$

7.5 Performance Analysis

7.5.1 Choice of the Laguerre and Kautz Network Parameters

If we chose N_s for a Laguerre network, then p will be automatically decided based on the fact that every Laguerre function should obey the orthonormality within the control horizon given by:

$$\sum_{m=0}^{N_c} l_i(m)l_j(m) = \begin{Bmatrix} 1, \text{ if } i = j \\ 0, \text{ if } i \neq j \end{Bmatrix} \tag{7.50}$$

In this book, N_s is chosen in such a way that it should be the minimum value for which the maximum error between the areas of all the input trajectories approximated with conventional pulse operators and Laguerre functions lies between $[-10^{-3}\ 10^{-3}]$ for the worst-case load disturbance scenario. The worst-case here is load hitting sudden maximum from the minimum level at all buses. The p.u minimum and maximum load parameters at different buses in the micro-grid shown in Fig. 3.1 of the Chap. 3 are given in APPENDIX 1. Based on this, the chosen Laguerre network parameters are $N_s = 50$ and $p = 0.5$.

If we chose the number of Kautz functions N_s in a Kautz network, then the pole parameters b, c will be decided based on the fact that every Kautz function should obey the orthonormality within the prediction horizon given by:

$$\sum_{m=0}^{N_c} K_i(m)K_j(m) = \begin{Bmatrix} 1, \text{ if } i = j \\ 0, \text{ if } i \neq j \end{Bmatrix} \tag{7.51}$$

Of course one of them (either b or c) is arbitrary. Then the other one is decided from (7.51). As already discussed above, N_s is chosen in such a way that it should be the minimum value for which the maximum error between the areas of all the input trajectories approximated with conventional pulse operators and Kautz functions lies between $[-10^{-3}\ 10^{-3}]$ for the worst-case load disturbance scenario. In this case, the chosen Kautz network parameters are $N_s = 50$, $b = 0.7984$, and $c = -0.2525$.

Example 7.1: LTI-MPC Performance with Special Functions
The performance analysis of the LTI-MPC with special functions is carried out for all the three cases of disturbances described in Chaps. 5 and 6. Figures 7.3, 7.4, and 7.5 show the performance of the LTI-MPC with special functions for the single R-L load disturbance, multiple R-L and IM load disturbances at a time, and PV-DG source intermittency. From the figures, we can observe that the performance of the LTI-MPC with special functions closely matches with the pulse operator-based LTI-MPC. However, the main advantage with the special function-based LTI-MPC is that at each sampling instant, the number of optimizing variables is 200 (4 inputs multiplied by 50 special functions for each input) when compared to the pulse operator-based LTI-MPC, where the number of optimizing variables are 600 (4 inputs multiplied by 150 pulses for each input). This has a considerable positive effect on the computational complexity of the controller when it is implemented online at the primary control level.

Example 7.2: LTV-MPC Performance with Special Functions
The performance analysis of the LTV-MPC with special functions in its formulation is carried out for all the three disturbances that are pointed out in Chaps. 5 and 6. Figures 7.6, 7.7, and 7.8 show the performance results. From the figures, we can observe that the performance of the LTV-MPC with special functions closely matches with the pulse operator-based LTV-MPC. However, the LTV-MPC with special functions requires only 200 optimizing variables at each sampling instant, whereas the pulse operator-based LTV-MPC require 600 optimizing variables.

7.6 Key Takeaways

- The chapter discussed the application of orthonormal special functions in the proposed MPC designs for the reduction of the online computational burden.
- The original optimal control trajectories within the control horizon are approximated with the orthonormal special function networks, namely Laguerre networks and two-parameter Kautz networks.
- The choice of the networks depends on the nature of the control trajectories that are to be approximated.
- Simulation results show that the use of orthonormal special functions in the LTI-MPC and LTV-MPC designs has drastically reduced the number of optimizing

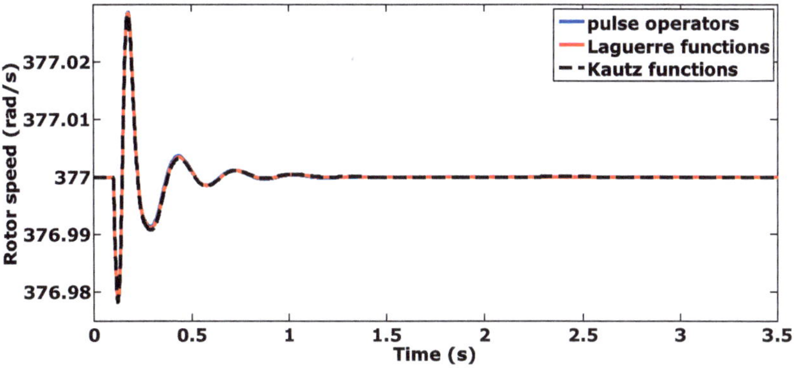

(a) Rotor speed during the single R-L load disturbance at bus B7

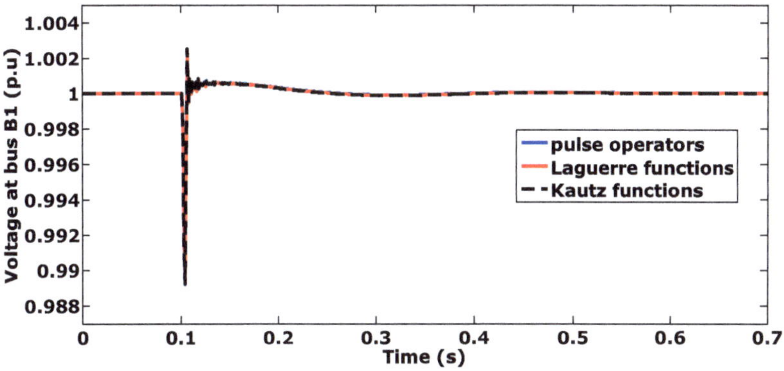

(b) Bus B1 voltage during the single R-L load disturbance at bus B7

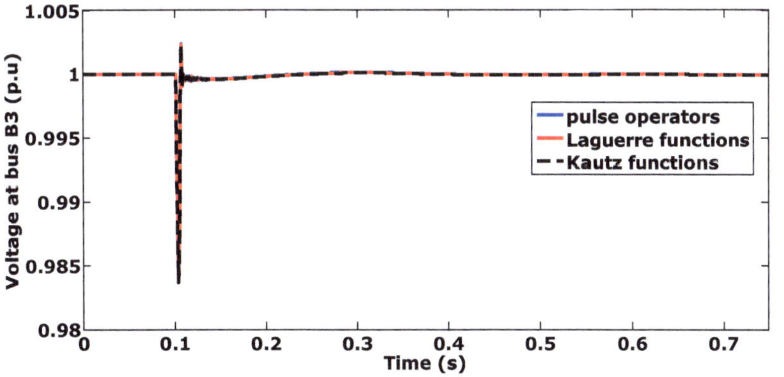

(c) Bus B3 voltage during the single R-L load disturbance at bus B7

Fig. 7.3 Performance of special function-based LTI-MPC for single R-L load disturbance: **a** rotor speed, **b** bus B1 voltage, and **c** bus B3 voltage

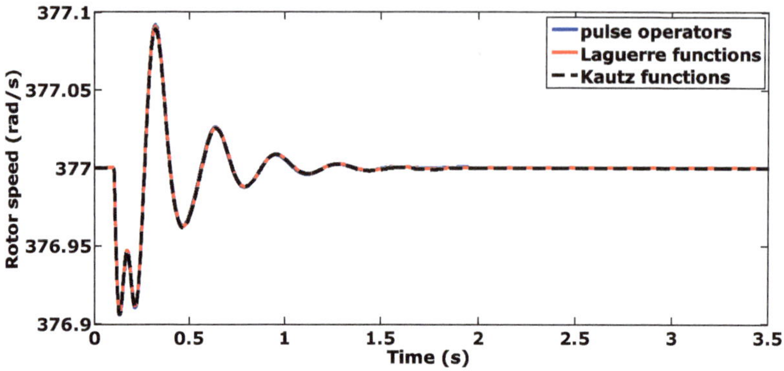

(a) Rotor speed during the multiple R-L and IM load disturbances at a time

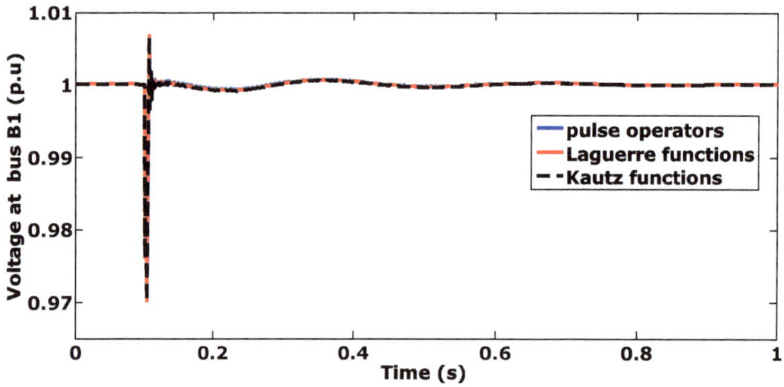

(b) Bus B1 voltage during the multiple R-L and IM load disturbances at a time

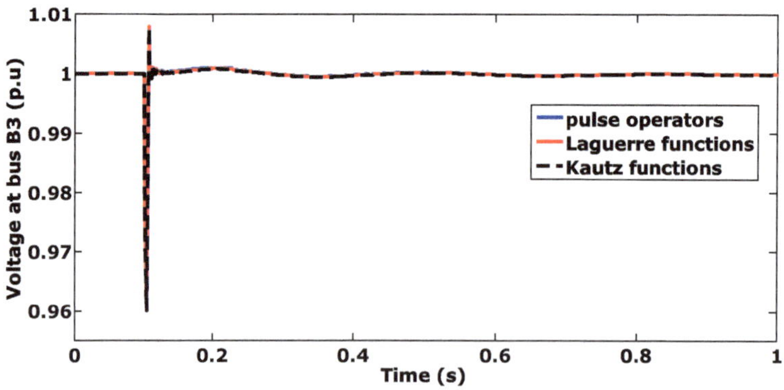

(c) B3 voltage during the multiple R-L and IM load disturbances at a time

Fig. 7.4 Performance of special function-based LTI-MPC for multiple R-L and IM load disturbances at a time: **a** rotor speed, **b** bus B1 voltage, and **c** bus B3 voltage

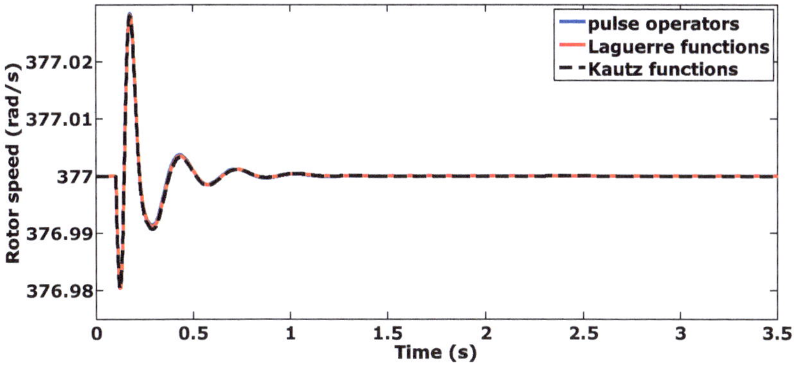

(a) Rotor speed during the PV-DG source intermittency

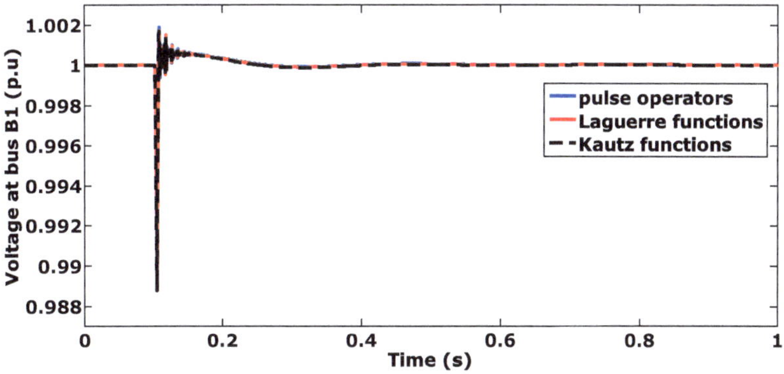

(b) Bus B1 voltage during the PV-DG source intermittency

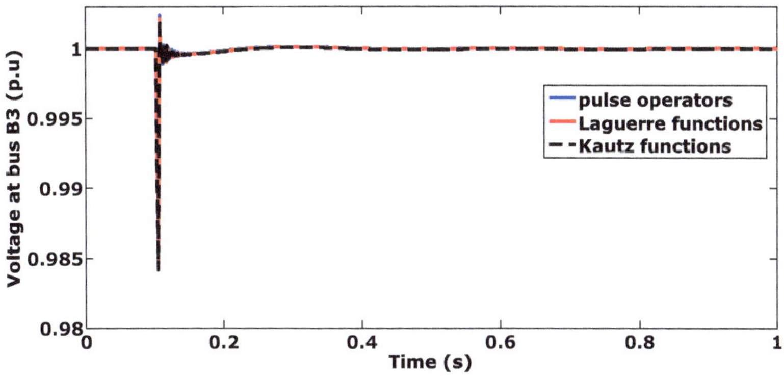

(c) Bus B3 voltage during the PV-DG source intermittency

Fig. 7.5 Performance of special function-based LTI-MPC for PV-DG source intermittency: **a** rotor speed, **b** bus B1 voltage, and **c** bus B3 voltage

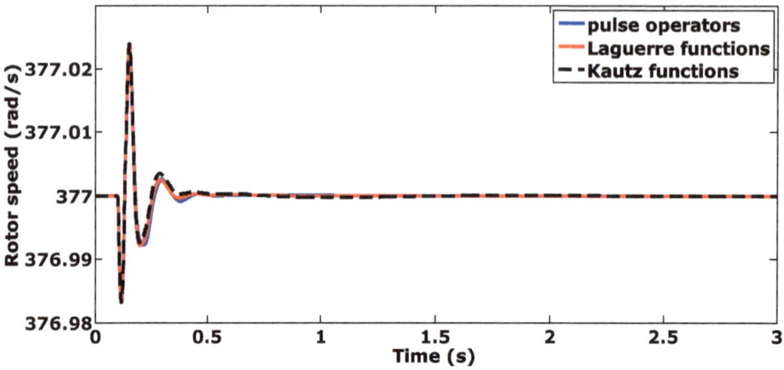

(a) Rotor speed during the single R-L load disturbance at bus B7

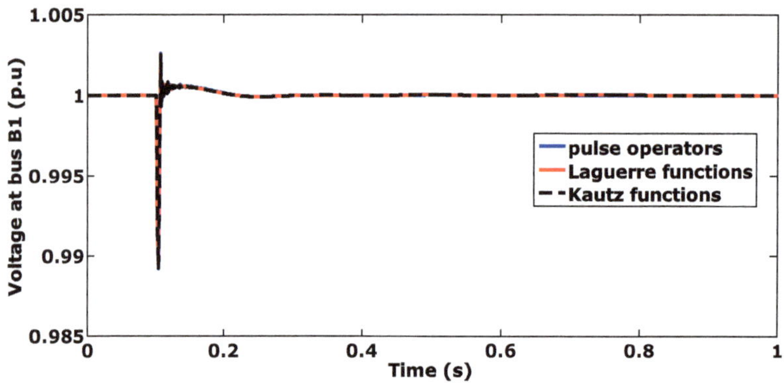

(b) Bus B1 voltage during the single R-L load disturbance at bus B7

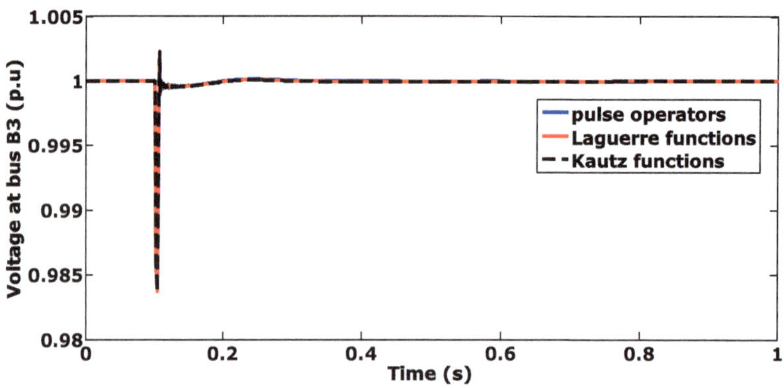

(c) Bus B3 voltage during the single R-L load disturbance at bus B7

Fig. 7.6 Performance of special functions based LTV-MPC for single R-L load disturbance: **a** rotor speed, **b** bus B1 voltage, and **c** bus B3 voltage

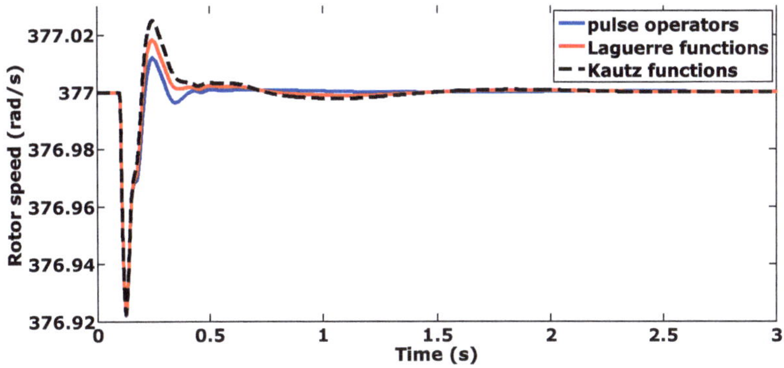

(a) Rotor speed during the multiple R-L and IM load disturbances at a time

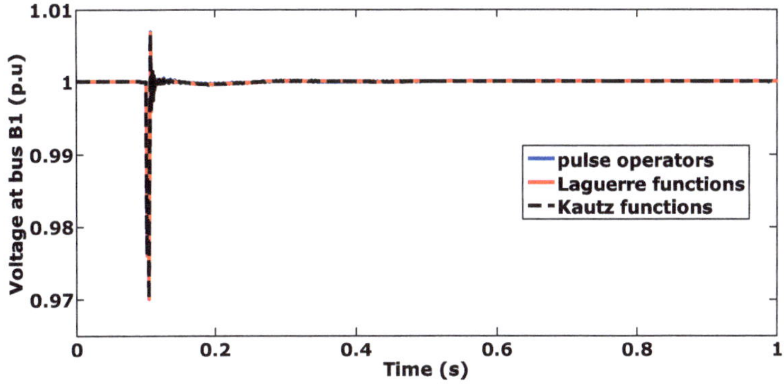

(b) Bus B1 voltage during the multiple R-L and IM load disturbances at a time

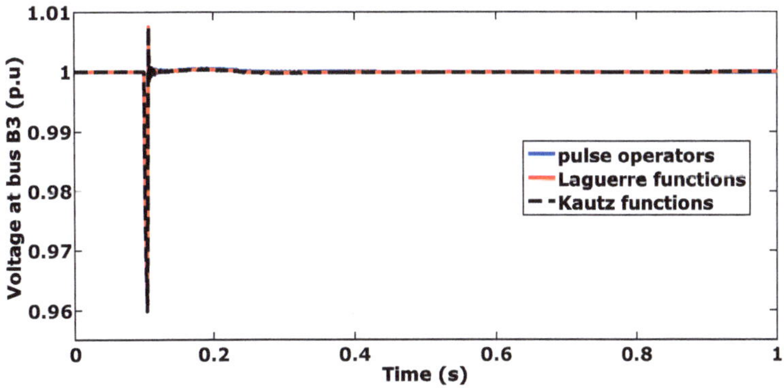

(c) Bus B3 voltage during the multiple R-L and IM load disturbances at a time

Fig. 7.7 Performance of special functions based LTV-MPC for multiple R-L and IM load disturbances at a time: **a** rotor speed, **b** bus B1 voltage, and **c** bus B3 voltage

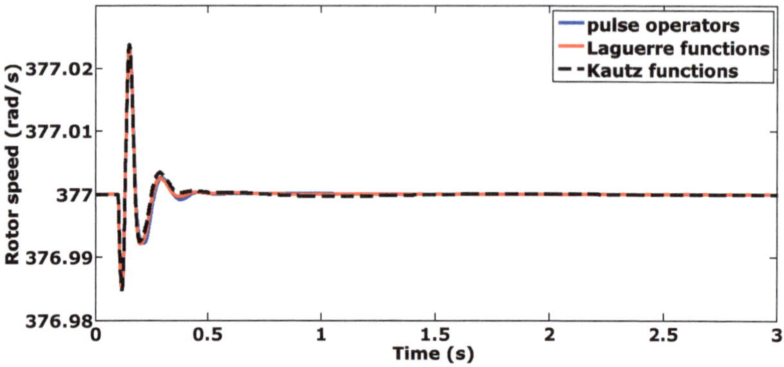

(a) Rotor speed during the PV-DG source intermittency

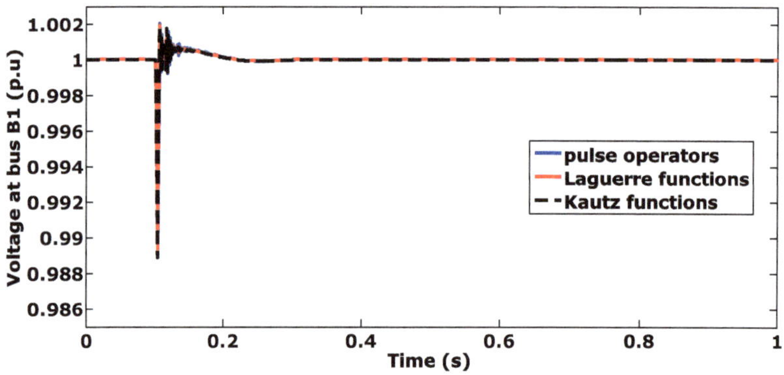

(b) Bus B1 voltage during the PV-DG source intermittency

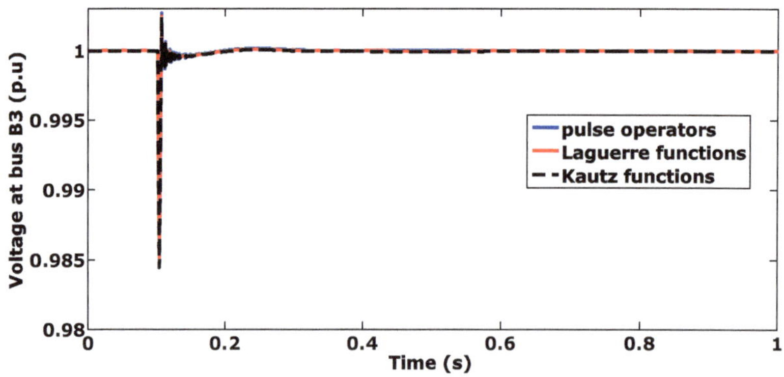

(c) Bus B3 voltage during the PV-DG source intermittency

Fig. 7.8 Performance of special functions based LTV-MPC for PV-DG source intermittency: **a** rotor speed, **b** bus B1 voltage, and **c** bus B3 voltage

variables in the optimal control problem without compromising the performance of the proposed MPC designs.
- This reduction in the number of optimizing variables improves the computational complexity of the respective MPC designs.

References

Cheng CM, Peng ZK, Dong XJ, Zhang WM, Meng G (2018) Nonlinear system identification using Kautz basis expansion-based Volterra–PARAFAC model. Nonlinear Dyn 94(3):2277–2287

Da Rosa A, Campello RJ, Amaral WC (2007) Choice of free parameters in expansions of discrete-time Volterra models using Kautz functions. Automatica 43(6):1084–1091

da Rosa A, Campello RJ, Amaral WC (2009) Exact search directions for optimization of linear and nonlinear models based on generalized orthonormal functions. IEEE Trans Autom Control 54(12):2757–2772

Heuberger PS, Van den Hof PM, Wahlberg B (eds) (2005) Modelling and identification with rational orthogonal basis functions. Springer Science & Business Media

Kautz WH (1954) Transient synthesis in the time domain. Trans IRE Professional Group Circuit Theor 3:29–39

Khan B, Rossiter JA, Valencia-Palomo G (2011) Exploiting Kautz functions to improve feasibility in MPC. IFAC Proc Vol 44(1):6777–6782

Misra S, Reddy R, Saha P (2016) Model predictive control of resonant systems using Kautz model. Int J Autom Comput 13(5):501–515

Puvvula V, Verma PP, Kesanakurthy SS (2020) Two-parameter Kautz network-based LTV-MPC for non-linear standalone micro-grid control. IET Renew Power Gener 14(12):2221–2231

Sai Sesha PSRVR, Kesanakurthy SS (2018) Model predictive control approach for frequency and voltage control of standalone micro-grid. IET Gener Transmission Distribution 12(14):3405–3413

Tanguy N, Morvan R, Vilbé P, Calvez LC (2002) Pertinent choice of parameters for discrete Kautz approximation. IEEE Trans Autom Control 47(5):783–787

Vidyasagar PS, Swamp KS (2016) Laguerre functions based model predictive frequency and voltage control of isolated micro-grid. In: 2016 IEEE Annual India Conference (INDICON). IEEE, pp 1–6

Vidyasagar P, Anoop VE, Swarup KS (2020) Application of continuous-time Kautz functions in the MPC formulation for standalone micro-grid control. In: 2020 21st National Power Systems Conference (NPSC). IEEE, pp 1–6

Wahlberg B (1994) System identification using Kautz modcls. IEEE Trans Autom Control 39(6):1276–1282

Wang L (2009) Model predictive control system design and implementation using MATLAB®. Springer Science & Business Media

Wang R (2012) Introduction to orthogonal transforms: with applications in data processing and analysis. Cambridge University Press, Cambridge

Chapter 8
Auxiliary Requirements for Real-Time Implementation

Abstract So far the book discussed the theoretical aspects of the MPC formulations for the micro-grid control at the primary control level. However, it is necessary to discuss some of the important requirements, referred to as the auxiliary requirements for the effective online implementation of the MPC formulations. The requirements comprise:

- Scalability
- Harmonics
- State estimation
- Choice of a particular MPC formulation
  - Based on computational complexity
  - Based on the performance comparison
- Robustness.

This chapter puts more focus on the choice among the MPC formulations and the incorporation of the robust features into the MPC formulations. The other three requirements, namely scalability, harmonics, and state estimation are briefly discussed in this chapter.

Keywords Computational complexity · Disturbance compensator · Harmonics · Scalability · State estimation · Robustness

8.1 Scalability

The centralized LTI-MPC and LTV-MPC discussed in the book can be implemented online for the primary control of micro-grids feeding small industrial corridors, commercial complexes, housing communities, and villages covering small geographical areas. Depending on the number of micro-grid inputs and the communication requirements, the sampling time of the controller can be adjusted from a millisecond to a few milliseconds in these micro-grids. While choosing the sampling time, the computational time of the optimal control trajectories, communication delay, and

P. Vidyasagar and K. Shanti Swarup, *Design and Development of Model Predictive Primary Control of Micro Grids*, Springer Tracts in Electrical and Electronics Engineering, https://doi.org/10.1007/978-981-19-5852-6_8

state estimation plays a key role. However, when the micro-grid is too large in terms of its geographical area or the number of generators in it, the control methodologies and the MPC formulations discussed in the book can still be applicable. But the implementation needs the following adjustments:

- Entire micro-grid network can be divided into small control centres that consist of a group of generators and loads. Each control centre is equipped with the centralized model predictive controller (MPC) formulations discussed in the book. Using the mathematical model and state information of the network under its control, MPC tries to control the frequency and voltage, as discussed in the previous chapters. However, in this case, a provision for cooperative control among different control centres can also be incorporated for effective power-sharing among the control centres. This type of implementation yields a better response than the one that is completely decentralized in which each generator is equipped with its own MPC (Yang et al. 2016; Maestre and Negenborn 2014).
- Another way is to completely decentralize the entire micro-grid. Every generator is having its own dedicated MPC. This cannot come under a centralized MPC scheme. However, the control methodologies and MPC formulations discussed in the book can be applied to each of the MPC's in the decentralized control (Tavakoli et al. 2016).

8.2 Harmonics

The PV-DG inverter is responsible for the injection of the harmonics into the micro-grid. However, only the fundamental components of the micro-grid mathematical model are considered in this book for the MPC formulations. This fundamental component consideration is based on the assumption that the harmonics can be taken care of by the selective harmonic elimination (SHE) module in the PWM generator of the PV-DG. The MPC formulations in this book are intended for the primary control of the fundamental components of the micro-grid frequency and generator bus voltages. At each sampling instant, the MPC formulations generate the optimal reference values for E_{d3} and E_{q3}. These are the fundamental $d3$–$q3$ components of the PV-DG inverter output voltage. These reference values are inputs to the PWM generator, which then calculates the fundamental modulation index (Tibola et al. 2011). The PV-DG inverter is equipped with the SHE module that consists of offline calculations of converter switching angles to eliminate the undesirable harmonics for different fundamental modulation indexes. These angles are calculated through the equations obtained from the Fourier analysis of different PWM patterns. These angles are stored in a look-up table inside of the SHE module, which is then read in real-time according to the desired fundamental modulation index. Higher-order harmonics that are not eliminated by the SHE module can be attenuated using the passive LC filter design of the PV-DG. Hence, only the fundamental model of the micro-grid is considered for the MPC formulations in this book.

8.3 State Estimation

In the MPC formulations, we assumed that the complete state information at the present sampling instant is available to us. However, in real-time, it is not possible to measure all the micro-grid states. In such cases, we need to estimate the states that are not measurable. To estimate these unknown states, observers are used in the MPC formulations. The tuning of observers can be carried out through pole assignment strategies or Kalman filters. In the combined closed-loop system of micro-grid with MPC and state observers, it is possible to design the MPC and state observers independently (Wang 2009). This particular conclusion allows us to completely focus on the MPC formulations without worrying about the impact of state observer dynamics on the MPC formulations.

8.4 Choice of a Particular MPC Formulation

One of the important auxiliary requirements is to find out which of the MPC formulations is suitable for the online implementation in a given micro-grid operational scenario. There are two ways to decide this. One way is from the computational complexity point of view, and the other way is from the superior performance point of view. The MPC formulation, which is superior in performance, is computationally complex and vice versa. Hence there is a trade-off between the performance and computational complexity while choosing a particular MPC formulation for a given sampling time of the controller.

8.4.1 Computational Complexity

For the computational complexity analysis let us consider a complexity factor 'C' which is given by:

$$\text{Complexity factor } (C) = \frac{\text{actual case time} - \text{base case time}}{\text{base case time}} \tag{8.1}$$

The case time here represents the time required to execute the complete operational procedure of the MPC formulations for one sample of time. The base case information was provided in the Appendix 1. The base case represents a prediction and control horizons of lengths $N_p = N_c = 100$ samples, one SG-DG ($N_{\text{sg}} = 1$), and one PV-DG ($N_{\text{pv}} = 1$). The penalty on each SG-DG input is taken to be 0.0000001, and on each PV-DG input is considered to be 0.0001. The sampling time is chosen to be $T_s = 1$ ms. The Laguerre network parameters are $N_s = 50$ and $p = 0.3$. The Kautz network parameters are $N_s = 50$, $b = 0.5492$, and $c = -0.0925$. For the single R-L load disturbance at bus B7 of the micro-grid shown in Fig. 3.1, the complexity factor

C is shown in Fig. 8.1, calculated for different prediction (control) horizons and a different number of synchronous and photovoltaic generators. An Intel® Core™ i3-5005U CPU@2.00 GHz processor is used for calculating the complexity factor C. From Fig. 8.1, we can observe that the LTI-MPC has less computational complexity compared to the LTV-MPC. In both cases (LTI and LTV), MPC formulations with special functions are computationally less complex compared to the pulse operator-based MPC formulations. Overall, LTI-MPC with special functions has less computational complexity, and LTV-MPC with pulse operators is more computationally complex.

8.4.2 Performance Point of View

It is also important to look from the performance point of view while choosing a particular MPC formulation. As we have already discussed in the previous chapters, the LTV-MPC performance is superior to the LTI-MPC performance. The special functions cannot improve the performance of the MPC formulations but can only match with that of the pulse operator-based MPC formulations. Hence LTV-MPC with pulse operators is superior in performance, whereas LTI-MPC with special functions is inferior in performance. A trade-off between the computational complexity and performance is necessary while choosing a particular MPC formulation for online implementation.

8.5 Robustness

One of the significant hurdles in the implementation of the online MPC is the robustness of the controller. MPC is highly dependent on the mathematical model of the system and the state information. Hence it is prone to steady-state errors and stability issues due to the parametric uncertainties, errors in the state information, and other unknown disturbances (Mayne et al. 2005; Chisci et al. 2001; Bemporad and Morari 1999). Hence there is a need for incorporation of robustness features into the MPC formulations. To achieve the required robustness, a disturbance compensator is used in this chapter (Merabet et al. 2017).

8.5.1 Disturbance Compensator

Considering the disturbances in the system, the complete mathematical model of the nonlinear micro-grid can be expressed by a state-space model given by:

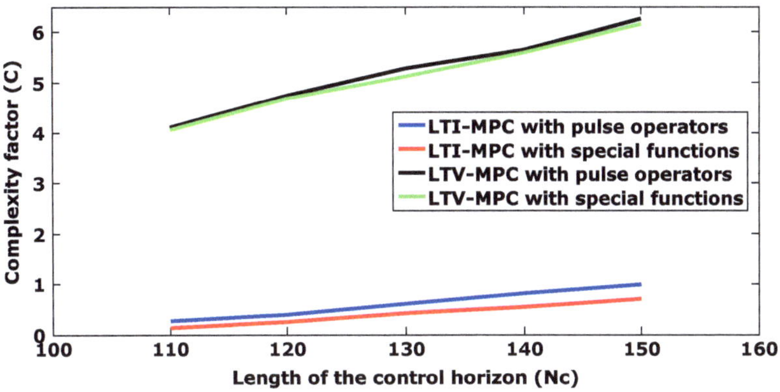

(a) Complexity factor for different control horizon lengths

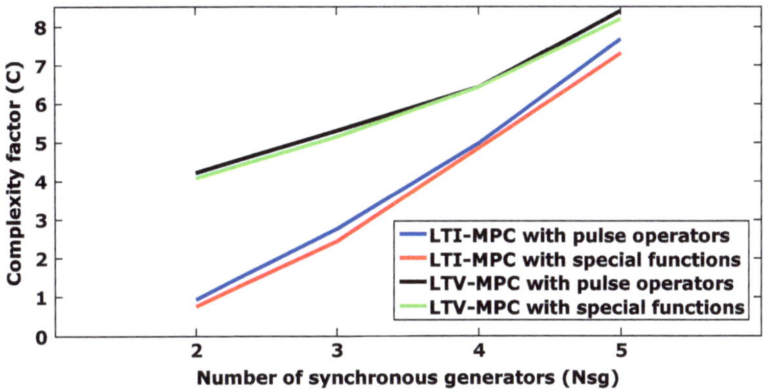

(b) Complexity factor for different number of synchronous generators

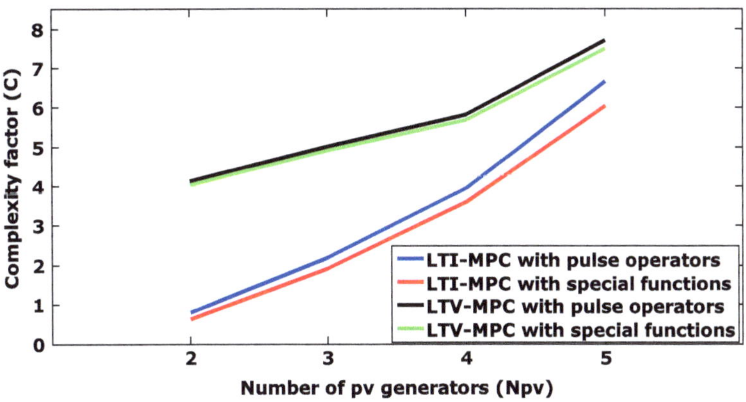

(c) Complexity factor for different number of PV generators

Fig. 8.1 Complexity factor comparison for single R-L load disturbance at bus B7: **a** control horizon lengths, **b** number of synchronous generators, and **c** number of PV generators

$$\frac{\mathrm{d}\boldsymbol{X}(t)}{\mathrm{d}t} = \boldsymbol{f}(\boldsymbol{X}(t), \boldsymbol{V}(t)) + \boldsymbol{\Upsilon}(t)$$
$$\boldsymbol{Y}(t) = \boldsymbol{g}(\boldsymbol{X}(t)) \tag{8.2}$$

$\boldsymbol{X}$ is the continuous-time state vector with n states, $\boldsymbol{V}$ is the input vector with n_{ip} inputs, and $\boldsymbol{Y}$ is the output vector with n_{op} outputs. $\boldsymbol{\Upsilon}$ is the disturbance vector that includes the parametric uncertainties, errors in the state information, and other unknown disturbances. The optimal control problem of the MPC formulations requires information about the vector $\boldsymbol{\Upsilon}$. However, as $\boldsymbol{\Upsilon}$ arises from the uncertainties in the system and unknown disturbances, we can only estimate it using the disturbance compensator. The disturbance compensator in continuous-time is given by the following mathematical model:

$$\frac{\mathrm{d}\widehat{\boldsymbol{\Upsilon}}(t)}{\mathrm{d}t} = \boldsymbol{\alpha}\left(\frac{\mathrm{d}\boldsymbol{X}(t)}{\mathrm{d}t} - \boldsymbol{f}(\boldsymbol{X}(t), \boldsymbol{V}(t))\right) - \boldsymbol{\alpha}\widehat{\boldsymbol{\Upsilon}}(t) \tag{8.3}$$

$\widehat{\boldsymbol{\Upsilon}}$ is the estimated vector of $\boldsymbol{\Upsilon}$ from the compensator. $\boldsymbol{\alpha}$ is the gain matrix of the compensator. Once the disturbance compensator model (8.3) is available, then it has to be integrated with the MPC formulations discussed in the previous chapters. In this chapter, the disturbance compensator model is integrated with the LTI-MPC formulations discussed in Chaps. 5 and 7. The same procedure can be adopted for the integration of the disturbance compensator model with the LTV-MPC formulations discussed in Chaps. 6 and 7.

8.5.2 Mathematical Formulation of the Robust LTI-MPC

The disturbance compensator gives the estimated disturbance vector $\widehat{\boldsymbol{\Upsilon}}$. For the formulation of the robust LTI-MPC, the nonlinear model of the micro-grid can be expressed as follows after considering the estimated disturbance vector $\widehat{\boldsymbol{\Upsilon}}$:

$$\frac{\mathrm{d}\boldsymbol{X}(t)}{\mathrm{d}t} = \boldsymbol{f}(\boldsymbol{X}(t), \boldsymbol{V}(t)) + \widehat{\boldsymbol{\Upsilon}}(t)$$
$$\boldsymbol{Y}(t) = \boldsymbol{g}(\boldsymbol{X}(t)) \tag{8.4}$$

The linear approximation of the continuous-time nonlinear model (8.4) within the prediction horizon is given by Taylor's series expansion with the higher-order terms neglected around the present operating point $(\boldsymbol{X}(t_s),\boldsymbol{V}(t_s-T_s))$. T_s is the sampling time of the controller.

$$\frac{d(\Delta \boldsymbol{X}(t))}{dt} = \boldsymbol{f}(\boldsymbol{X}(t_s), \boldsymbol{V}(t_s - T_s)) + \frac{\partial \boldsymbol{f}}{\partial \boldsymbol{X}}\bigg|_{\substack{\boldsymbol{X} = \boldsymbol{X}(t_s) \\ \boldsymbol{V} = \boldsymbol{V}(t_s - T_s)}} \Delta \boldsymbol{X}(t)$$

$$+\frac{\partial \boldsymbol{f}}{\partial \boldsymbol{V}}\Big|_{\substack{\boldsymbol{X}=\boldsymbol{X}(t_s)\\ \boldsymbol{V}=\boldsymbol{V}(t_s-T_s)}} \Delta \boldsymbol{V}(t)+\widehat{\boldsymbol{\Upsilon}}(t_s)$$

$$\boldsymbol{Y}(t)=\boldsymbol{g}(\boldsymbol{X}(t_s))+\frac{\partial \boldsymbol{g}}{\partial \boldsymbol{X}}\Big|_{\substack{\boldsymbol{X}=\boldsymbol{X}(t_s)\\ \boldsymbol{V}=\boldsymbol{V}(t_s-T_s)}} \Delta \boldsymbol{X}(t)$$

$$\Delta \boldsymbol{V}(t)=\boldsymbol{V}(t)-\boldsymbol{V}(t_s-T_s),\ \Delta \boldsymbol{X}(t)=\boldsymbol{X}(t)-\boldsymbol{X}(t_s) \tag{8.5}$$

It is assumed that the external disturbance within the prediction horizon is constant and is equal to its value at the present time $t = t_s$ seconds. The reason for this assumption is that it is difficult to predict the unknown disturbances within the prediction horizon.

$$\widehat{\boldsymbol{\Upsilon}}(t)=\widehat{\boldsymbol{\Upsilon}}(t_s),\ \left(t_s \le t \le t_s+N_p T_s\right) \tag{8.6}$$

Even though this assumption is not accurate, due to the receding horizon principle of the MPC, the error due to this assumption can be negotiated. The linear model (8.6) can be written as:

$$\begin{aligned}\frac{\mathrm{d}\boldsymbol{Z}_c(t)}{\mathrm{d}t}&=\overline{\boldsymbol{Z}}+\overline{\boldsymbol{\Upsilon}}+\boldsymbol{A}_t\boldsymbol{Z}_c(t)+\boldsymbol{B}_t\boldsymbol{u}(t)\\ \boldsymbol{Y}(t)&=\overline{\boldsymbol{Y}}+\boldsymbol{C}_t\boldsymbol{Z}_c(t)\end{aligned} \tag{8.7}$$

$$\boldsymbol{A}_t=\frac{\partial \boldsymbol{f}}{\partial \boldsymbol{X}}\Big|_{\substack{X=X(t_s)\\ V=V(t_s-T_s)}},\ \boldsymbol{B}_t=\frac{\partial \boldsymbol{f}}{\partial \boldsymbol{V}}\Big|_{\substack{X=X(t_s)\\ V=V(t_s-T_s)}},$$

$$\boldsymbol{C}_t=\frac{\partial \boldsymbol{g}}{\partial \boldsymbol{X}}\Big|_{\substack{X=X(t_s)\\ V=V(t_s-T_s)}},\ \boldsymbol{Z}_c=\Delta\boldsymbol{X},\ \boldsymbol{u}=\Delta\boldsymbol{V},\ \overline{\boldsymbol{\Upsilon}}=\widehat{\boldsymbol{\Upsilon}}(t_s),$$

$$\overline{\boldsymbol{Z}}=\boldsymbol{f}(\boldsymbol{X}(t_s),\boldsymbol{V}(t_s-T_s)),\ \overline{\boldsymbol{Y}}=\boldsymbol{g}(\boldsymbol{X}(t_s))$$

The continuous model (8.7) is then converted to a discrete model using zero-order hold (ZOH) given by:

$$\begin{aligned}\boldsymbol{Z}_c(k+1)&=\widetilde{\boldsymbol{Z}}+\widetilde{\boldsymbol{\Upsilon}}+\boldsymbol{A}_d\boldsymbol{Z}_c(k)+\boldsymbol{B}_d\boldsymbol{u}(k)\\ \boldsymbol{Y}(k)&=\widetilde{\boldsymbol{Y}}+\boldsymbol{C}_d\boldsymbol{Z}_c(k)\end{aligned} \tag{8.8}$$

$(\boldsymbol{A}_d, \boldsymbol{B}_d, \boldsymbol{C}_d)$ is the discrete-time state-space triplet with $\{k = 0 \Leftrightarrow t = t_s\}$. $\widetilde{\boldsymbol{Z}}$, $\widetilde{\boldsymbol{Y}}$, $\widetilde{\boldsymbol{\Upsilon}}$ are discrete-time counterparts to $\overline{\boldsymbol{Z}}, \overline{\boldsymbol{Y}}, \overline{\boldsymbol{\Upsilon}}$. As the micro-grid control requires an integral action, integrators are embedded in (8.8) by converting it to an augmented model:

$$\begin{aligned} \boldsymbol{Z}(k+1) &= \overline{\boldsymbol{\theta}} + \overline{\boldsymbol{\xi}} + \boldsymbol{A}\boldsymbol{Z}(k) + \boldsymbol{B}\Delta\boldsymbol{u}(k) \\ \boldsymbol{Y}(k) &= \widetilde{\boldsymbol{Y}} + \boldsymbol{C}\boldsymbol{Z}(k) \end{aligned} \tag{8.9}$$

$$\boldsymbol{A} = \begin{bmatrix} \boldsymbol{A_d} & \boldsymbol{B_d} \\ 0_{n_{ip}\times n} & \boldsymbol{I}_{n_{ip}\times n_{ip}} \end{bmatrix}, \boldsymbol{B} = \begin{bmatrix} \boldsymbol{B_d} \\ \boldsymbol{I}_{n_{ip}\times n_{ip}} \end{bmatrix}, \boldsymbol{C} = \begin{bmatrix} \boldsymbol{C_d} & 0_{n_{op}\times n_{ip}} \end{bmatrix}, \overline{\boldsymbol{\theta}}^{\mathrm{T}} = \begin{bmatrix} \widetilde{\boldsymbol{Z}}^{\mathrm{T}} & 0_{1\times n_{ip}} \end{bmatrix},$$

$$\boldsymbol{Z}(k)^{\mathrm{T}} = \begin{bmatrix} \boldsymbol{Z_c}(k)^{\mathrm{T}} & \boldsymbol{u}(k-1)^{\mathrm{T}} \end{bmatrix}, \overline{\boldsymbol{\xi}}^{\mathrm{T}} = \begin{bmatrix} \widetilde{\boldsymbol{\Upsilon}}^{\mathrm{T}} & 0_{1\times n_{ip}} \end{bmatrix}$$

Let $\boldsymbol{U}$ be a vector that contains the future control trajectories from the present sampling instant $k = k_i$ to a control horizon of length N_c samples. Let $\boldsymbol{Y_E}$ be a vector with micro-grid outputs predicted from the sampling instant $k = k_i + 1$ to a prediction horizon of length N_p samples.

$$\boldsymbol{Y_E} = \begin{bmatrix} \boldsymbol{Y}(k_i+1)^{\mathrm{T}} & \boldsymbol{Y}(k_i+2)^{\mathrm{T}} \ldots \boldsymbol{Y}(k_i+N_P)^{\mathrm{T}} \end{bmatrix}^{\mathrm{T}} \tag{8.10}$$

$$\boldsymbol{U} = \begin{bmatrix} \Delta\boldsymbol{u}(k_i)^{\mathrm{T}} & \Delta\boldsymbol{u}(k_i+1)^{\mathrm{T}} \ldots \Delta\boldsymbol{u}(k_i+N_C-1)^{\mathrm{T}} \end{bmatrix}^{\mathrm{T}} \tag{8.11}$$

From (8.9), we can write the following equation:

$$\begin{aligned} \boldsymbol{Y}(k_i+m) = \widetilde{\boldsymbol{Y}} + \boldsymbol{C}\boldsymbol{A}^m\boldsymbol{Z}(k_i) + \boldsymbol{C}\sum_{j=0}^{m-1}\boldsymbol{A}^j\overline{\boldsymbol{\theta}} + \boldsymbol{C}\sum_{j=0}^{m-1}\boldsymbol{A}^j\overline{\boldsymbol{\xi}} \\ + \boldsymbol{C}\sum_{j=0}^{m-1}\boldsymbol{A}^{m-1-j}\boldsymbol{B}\Delta\boldsymbol{u}(k_i+j) \end{aligned} \tag{8.12}$$

Substituting (8.12) in (8.10), we can get the following equation:

$$\boldsymbol{Y_E} = \boldsymbol{Y}_1 + \boldsymbol{Y}_2\overline{\boldsymbol{\theta}} + \boldsymbol{Y}_3\boldsymbol{Z}(k_i) + \boldsymbol{Y}_4\boldsymbol{U} + \boldsymbol{Y}_5\overline{\boldsymbol{\xi}} \tag{8.13}$$

$$\boldsymbol{Y}_1 = \begin{bmatrix} \widetilde{\boldsymbol{Y}}^{\mathrm{T}} & \widetilde{\boldsymbol{Y}}^{\mathrm{T}} & \widetilde{\boldsymbol{Y}}^{\mathrm{T}} \ldots \widetilde{\boldsymbol{Y}}^{\mathrm{T}} \end{bmatrix}^{\mathrm{T}}, \boldsymbol{Y}_2 = \begin{bmatrix} \boldsymbol{C}^{\mathrm{T}} & (\boldsymbol{C}+\boldsymbol{C}\boldsymbol{A})^{\mathrm{T}} \ldots \left(\boldsymbol{C}+\boldsymbol{C}\boldsymbol{A}+\cdots+\boldsymbol{C}\boldsymbol{A}^{N_p-1}\right)^{\mathrm{T}} \end{bmatrix}^{\mathrm{T}},$$

$$\boldsymbol{Y}_3 = \begin{bmatrix} (\boldsymbol{C}\boldsymbol{A})^{\mathrm{T}} & \left(\boldsymbol{C}\boldsymbol{A}^2\right)^{\mathrm{T}} \ldots \left(\boldsymbol{C}\boldsymbol{A}^{N_p}\right)^{\mathrm{T}} \end{bmatrix}^{\mathrm{T}}, \boldsymbol{Y}_4 = \begin{bmatrix} \boldsymbol{F}_1^{\mathrm{T}} & \boldsymbol{F}_2^{\mathrm{T}} & \boldsymbol{F}_3^{\mathrm{T}} \ldots \boldsymbol{F}_{N_p}^{\mathrm{T}} \end{bmatrix}^{\mathrm{T}},$$

$$\boldsymbol{F}_j = \begin{bmatrix} \boldsymbol{C}\boldsymbol{A}_{j,1}\boldsymbol{B} & \boldsymbol{C}\boldsymbol{A}_{j,2}\boldsymbol{B} & \boldsymbol{C}\boldsymbol{A}_{j,3}\boldsymbol{B} \ldots \boldsymbol{C}\boldsymbol{A}_{j,N_p}\boldsymbol{B} \end{bmatrix}, \boldsymbol{A}_{j,l} = \begin{Bmatrix} \boldsymbol{A}^{j-l} & \text{if} j \geq l \\ 0 & \text{if} j < l \end{Bmatrix},$$

$$\boldsymbol{Y}_5 = \begin{bmatrix} \boldsymbol{C}^{\mathrm{T}} & (\boldsymbol{C}+\boldsymbol{C}\boldsymbol{A})^{\mathrm{T}} \ldots \left(\boldsymbol{C}+\boldsymbol{C}\boldsymbol{A}+\cdots+\boldsymbol{C}\boldsymbol{A}^{N_p-1}\right)^{\mathrm{T}} \end{bmatrix}^{\mathrm{T}}$$

Once $\boldsymbol{Y_E}$ is formed from (8.13), an objective function is formulated using $\boldsymbol{Y_E}$ and is evaluated (minimized) subjected to a set of constraints on the micro-grid operation to generate the optimal control trajectories given by $\boldsymbol{U} = \boldsymbol{U}_{\mathbf{opt}}$.

$$\text{Min } J = (\boldsymbol{W} - \boldsymbol{Y_E})^T(\boldsymbol{W} - \boldsymbol{Y_E}) + \boldsymbol{U}^T\boldsymbol{R}\boldsymbol{U} \tag{8.14}$$

$$\text{S. T.}\Delta \boldsymbol{u}_{\mathbf{min}} \leq \Delta \boldsymbol{u}(k_i) \leq \Delta \boldsymbol{u}_{\mathbf{max}}, \boldsymbol{u}_{\mathbf{min}} \leq \boldsymbol{u}(k_i) \leq \boldsymbol{u}_{\mathbf{max}}, \boldsymbol{V}_{\mathbf{min}} \leq \boldsymbol{V}(k_i) \leq \boldsymbol{V}_{\mathbf{max}},$$

$$P_{G1,\min} \leq P_{G1}(k_i + 1) \leq P_{G1,\max}, Q_{G1,\min} \leq Q_{G1}(k_i + 1) \leq Q_{G1,\max},$$

$$P_{G3,\min} \leq P_{G3}(k_i + 1) \leq P_{G3,\max}, Q_{G3,\min} \leq Q_{G3}(k_i + 1) \leq Q_{G3,\max},$$

$$V_{\text{mppt}}\,(G, T) \leq V_{\text{dc}}(k_i + 1) \leq V_{\text{oc}}\,(G, T)$$

$\boldsymbol{W}$ is a vector with future set-points for the outputs. $\boldsymbol{R}$ is a positive definite weight matrix on input increments. The minimum and maximum limits of the active and reactive powers of the SG-DG and PV-DG are given by $\left(P_{G1,\min}, P_{G1,\max}\right),\left(Q_{G1,\min}, Q_{G1,\max}\right),\left(P_{G3,\min}, P_{G3,\max}\right),\left(Q_{G3,\min}, Q_{G3,\max}\right)$.. (P_{G1}, Q_{G1}), (P_{G3}, Q_{G3}) are active and reactive powers of the SG-DG and PV-DG. $V_{\text{mppt}}(G, T)$ and $V_{\text{oc}}\,(G, T)$ are the output voltages of the PV array corresponding to the maximum power point and open circuit. V_{dc} is the DC-link capacitor voltage.

8.5.3 *Mathematical Formulation of the Robust LTI-MPC with Special Functions*

Within the prediction horizon, the cost function (8.14) can be written as:

$$\begin{aligned} J = &\sum_{m=1}^{N_p} (\boldsymbol{w}(k_i) - \boldsymbol{Y}(k_i + m))^{\mathrm{T}} (\boldsymbol{w}(k_i) - \boldsymbol{Y}(k_i + m)) \\ &+ \sum_{m=0}^{N_c - 1} \Delta \boldsymbol{u}(k_i + m)^{\mathrm{T}} \boldsymbol{r}(k_i) \Delta \boldsymbol{u}(k_i + m) \end{aligned} \tag{8.15}$$

$\boldsymbol{w}(k_i)$ consists of set-points for each output in the output vector $\boldsymbol{Y}(k_i + m)$. $\boldsymbol{r}(k_i)$ consists of penalties on each input increment in the vector $\Delta \boldsymbol{u}(k_i + m)$. As discussed in Chap. 7, we can write the input vector $\Delta \boldsymbol{u}(k_i + m)$ of (8.11) as:

$$\Delta \boldsymbol{u}(k_i + m) = \boldsymbol{O}(m)^{\mathrm{T}} \boldsymbol{\eta}, \boldsymbol{O} = \boldsymbol{L}(\boldsymbol{or})\boldsymbol{K} \tag{8.16}$$

The details and formats of the $\boldsymbol{O}$ and $\boldsymbol{\eta}$ of the special functions can be understood from Chap. 7. Using (8.16), we can write the following equation:

$$\sum_{m=0}^{N_c - 1} \Delta \boldsymbol{u}(k_i + m)^{\mathrm{T}} \boldsymbol{r}(k_i) \Delta \boldsymbol{u}(k_i + m) = \sum_{m=0}^{N_c - 1} \boldsymbol{\eta}^{\mathrm{T}} \boldsymbol{O}(m) \boldsymbol{r}(k_i) \boldsymbol{O}(m)^{\mathrm{T}} \boldsymbol{\eta} \tag{8.17}$$

Since the special functions are orthonormal, we have:

$$\sum_{m=0}^{\infty} o_i(m) o_j(m) = \begin{Bmatrix} 1, & \text{if } i = j \\ 0, & \text{if } i \neq j \end{Bmatrix} \tag{8.18}$$

Applying orthonormal property (8.18) on (8.17) for sufficiently large control horizon length N_c, we can write the following:

$$\sum_{m=0}^{N_c-1} \Delta \boldsymbol{u}(k_i+m)^{\mathrm{T}} \boldsymbol{r}(k_i) \Delta \boldsymbol{u}(k_i+m) \cong \boldsymbol{\eta}^{\mathrm{T}} \boldsymbol{R_w} \boldsymbol{\eta} \tag{8.19}$$

$\boldsymbol{R_w}$ is the penalty matrix on input increments. Substituting (8.12), (8.16), and (8.19) in the quadratic cost function (8.15), we get the following convex quadratic optimal control problem:

$$\mathrm{Min} J = \boldsymbol{\eta}^{\mathrm{T}} \boldsymbol{E} \boldsymbol{\eta} + 2\boldsymbol{\eta}^{\mathrm{T}}(-\boldsymbol{E}_1 + \boldsymbol{E}_2 + \boldsymbol{E}_3 + \boldsymbol{E}_4 + \boldsymbol{E}_5) + \text{non } \boldsymbol{\eta} \text{ terms} \tag{8.20}$$

$$\text{S. T.} \Delta \boldsymbol{u}_{\min} \leq \Delta \boldsymbol{u}(k_i) \leq \Delta \boldsymbol{u}_{\max}, \boldsymbol{u}_{\min} \leq \boldsymbol{u}(k_i) \leq \boldsymbol{u}_{\max}, \boldsymbol{V}_{\min} \leq \boldsymbol{V}(k_i) \leq \boldsymbol{V}_{\max},$$

$$P_{G1,\min} \leq P_{G1}(k_i+1) \leq P_{G1,\max}, Q_{G1,\min} \leq Q_{G1}(k_i+1) \leq Q_{G1,\max},$$

$$P_{G3,\min} \leq P_{G3}(k_i+1) \leq P_{G3,\max}, Q_{G3,\min} \leq Q_{G3}(k_i+1) \leq Q_{G3,\max},$$

$$V_{\mathrm{mppt}}(G,T) \leq V_{\mathrm{dc}}(k_i+1) \leq V_{\mathrm{oc}}\,(G,T), \boldsymbol{E} = \boldsymbol{R_w} + \sum_{m=1}^{N_p} \phi(m) \boldsymbol{Q} \phi(m)^{\mathrm{T}}, \boldsymbol{Q} = \boldsymbol{C}^{\mathrm{T}} \boldsymbol{C},$$

$$\boldsymbol{E}_1 = \sum_{m=1}^{N_p} \phi(m) \boldsymbol{C}^{\mathrm{T}} \boldsymbol{w}(k_i), \boldsymbol{E}_2 = \sum_{m=1}^{N_p} \phi(m) \boldsymbol{C}^{\mathrm{T}} \widetilde{\boldsymbol{Y}}, \boldsymbol{E}_3 = \sum_{m=1}^{N_p} \phi(m) \boldsymbol{Q} \boldsymbol{F}(m) \overline{\boldsymbol{\theta}},$$

$$\boldsymbol{F}(m) = \sum_{j=0}^{m-1} \boldsymbol{A}^j, \boldsymbol{E}_4 = \sum_{m=1}^{N_p} \phi(m) \boldsymbol{Q} \boldsymbol{A}^m \boldsymbol{Z}(k_i), \boldsymbol{E}_5 = \sum_{m=1}^{N_p} \phi(m) \boldsymbol{Q} \boldsymbol{F}(m) \overline{\boldsymbol{\xi}}$$

$$\phi(m)^{\mathrm{T}} = \sum_{j=0}^{m-1} \boldsymbol{A}^{m-1-j} \boldsymbol{B} \boldsymbol{O}(j)^{\mathrm{T}}$$

$(P_{G1,\min}, P_{G1,\max})$, $(Q_{G1,\min}, Q_{G1,\max})$, $(P_{G3,\min}, P_{G3,\max})$, $(Q_{G3,\min}, Q_{G3,\max})$ are the minimum and maximum limits of the active and reactive powers of the SG-DG and PV-DG. $V_{\mathrm{mppt}}(G, T)$ and $V_{\mathrm{oc}}(G, T)$ are the output voltages of the PV array corresponding to the maximum power point and open circuit. The convex quadratic optimal control problem (8.20) is evaluated, and an optimal coefficient vector $\boldsymbol{\eta} =$

$\eta_{\mathbf{opt}}$ is generated. By applying the receding horizon principle, inputs corresponding to the present sampling instant $\Delta \boldsymbol{u}(k_i) = \boldsymbol{O}(0)^{\mathrm{T}} \boldsymbol{\eta}_{\mathbf{opt}}$ are calculated.

8.6 Performance Analysis of the Robust LTI-MPC

The performance of the robust LTI-MPC with a disturbance compensator is analysed for.

- Parametric uncertainties
- Errors in the state information
- Unknown disturbances in the micro-grid.

For the analysis, each gain in the compensator gain matrix $\boldsymbol{\alpha}$ is taken to be 500. This numerical value can arrive either through the stability analysis, or through the one at a time (OAT) sensitivity analysis. Figure 8.2 shows the performance of the robust LTI-MPC when there are parametric uncertainties in the micro-grid shown in Fig. 3.1. The parameters considered here are the stator resistance of the SG-DG (R_s), PV-DG filter parameters (R_f, X_f), PV-DG interfacing transformer parameters (R_t, X_t), line parameters between buses B1 and B4 (R_{14}, X_{14}), B2 and B4 (R_{24}, X_{24}), B3 and B4 (R_{34}, X_{34}). At $t = 0.1$ s, all these parameters are set to 1.2 times their actual values in the micro-grid model considered for the LTI-MPC formulation. The wrong information about these micro-grid parameters given in the micro-grid model results in inaccurate predictions of the micro-grid response within the prediction horizon. This further causes the frequency and voltage deviations in the micro-grid, as shown in Fig. 8.2. Without the disturbance observer, these deviations sustain in the system and may lead to either steady-state offsets or instability in the micro-grid. Figure 8.2 shows that the robustness in the LTI-MPC due to the disturbance compensator brings back the frequency and voltages to their nominal values when there are parametric uncertainties in the micro-grid. This indicates that the disturbance compensator accounts for these parametric uncertainties through the estimated disturbance vector $\widehat{\boldsymbol{\Upsilon}}$.

An error in the state information is one of the very common problems for the online implementation of the MPC. The errors generally occur from the bad design of the state estimator or from the bad measurements. Figure 8.3 shows the performance of the robust LTI-MPC when there are errors in the state information of the micro-grid. The states considered here are the damper winding flux linkages of the SG-DG (Ψ_{ad1}, Ψ_{aq1}), the intermediate state of PLL φ_{PLL}, load currents at buses B5 (I_{l5}^D, I_{l5}^Q) and B8 (I_{l8}^D, I_{l8}^Q). At $t = 0.1$ s, the information of these states is intentionally amplified to 1.2 times their original values while giving them to the LTI-MPC. Figure 8.3 shows that the robustness in the LTI-MPC due to the disturbance compensator brings back the frequency and voltages to their nominal values when there are errors in the state information of the micro-grid. The disturbance compensator accounts for these state information errors through the estimated disturbance vector $\widehat{\boldsymbol{\Upsilon}}$.

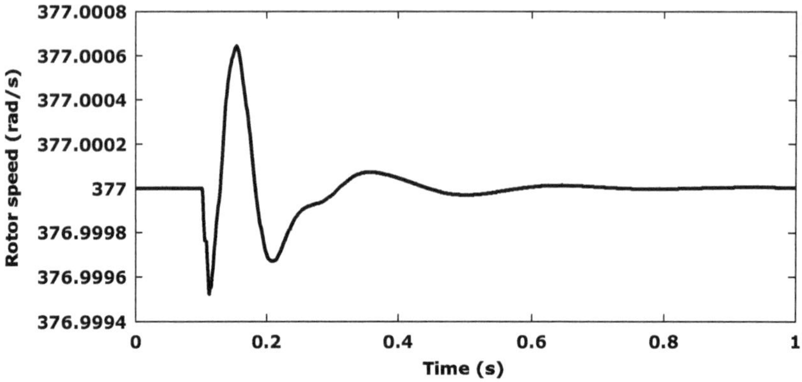

(a) Rotor speed during parametric uncertainties

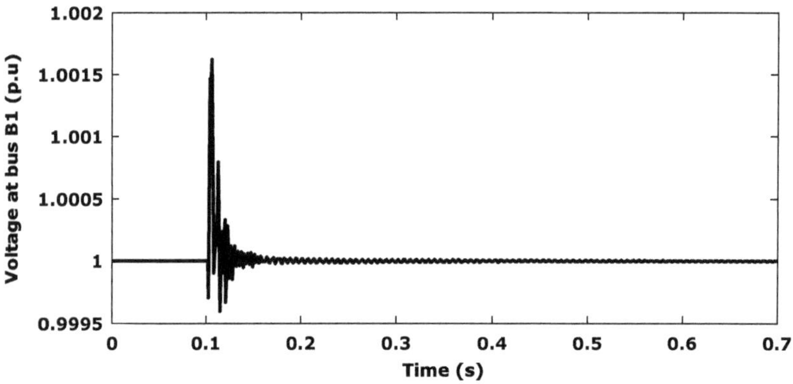

(b) Bus B1 voltage during parametric uncertainties

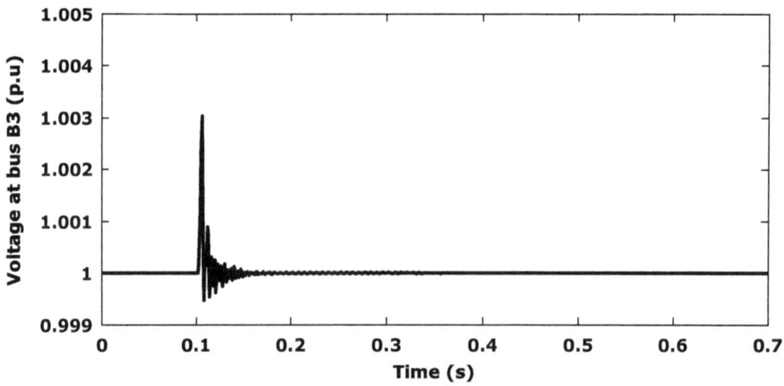

(c) Bus B3 voltage during parametric uncertainties

Fig. 8.2 Performance of the robust LTI-MPC for parametric uncertainties: **a** rotor speed, **b** bus B1 voltage, and **c** bus B3 voltage

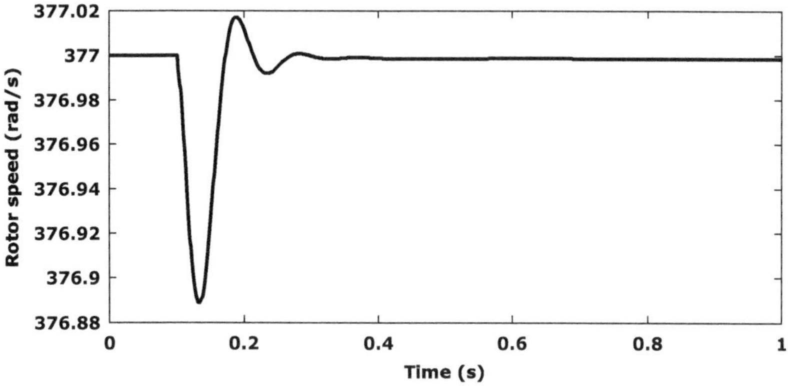

(a) Rotor speed during state information errors

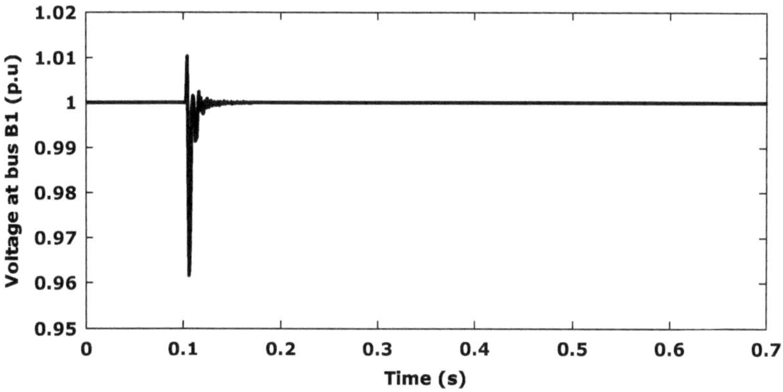

(b) Bus B1 voltage during state information errors

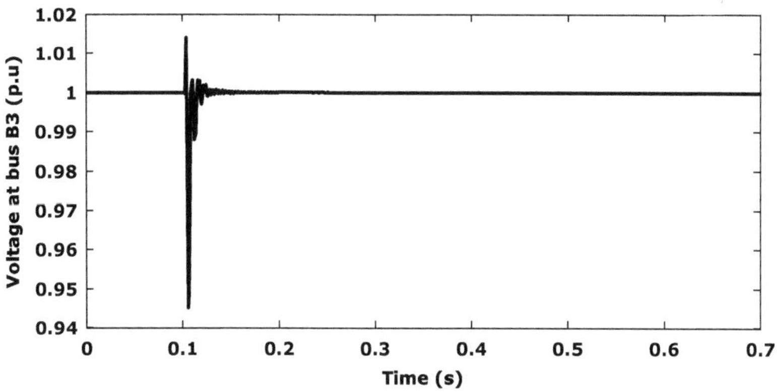

(c) Bus B3 voltage during state information errors

Fig. 8.3 Performance of the robust LTI-MPC for state information errors: **a** rotor speed, **b** bus B1 voltage, and **c** bus B3 voltage

Apart from state estimation errors and parametric uncertainties, the other challenge in the online implementation of the MPC arises from the unknown disturbances in the micro-grid. Figure 8.4 shows the performance of the robust LTI-MPC for constant disturbance values in the turbine dynamics (3.13), (3.14) of the SG-DG model given in Chap. 3. Each of these constant values is 0.1 p.u. Figure 8.4 shows that the robustness in the LTI-MPC due to the disturbance compensator brings back the frequency and voltages to their nominal values when there are unknown disturbances in the micro-grid. The disturbance compensator accounts for these unknown disturbances through the estimated disturbance vector $\widehat{\boldsymbol{\Upsilon}}$.

8.7 Key Takeaways

- The chapter solely focused on the supportive requirements for the online implementation of the MPC formulations.
- The chapter explained the scalability of the MPC formulations and concluded that the proposals could be extended to large scale micro-grids with a cooperative MPC structure.
- The selective harmonic eliminator (SHE) and state estimator can be designed independently of the MPC formulations.
- The chapter discussed how to choose a particular MPC formulation. The choice depends on a trade-off between computational complexity and performance.
- Simulation results show that the LTI-MPC with special functions is least complex, whereas LTV-MPC with special functions has superior performance.
- Finally, the chapter discussed the integration of the disturbance compensator with the proposed MPC formulations so that the overall MPC becomes robust to parametric uncertainties, state estimation errors, and unknown disturbances.

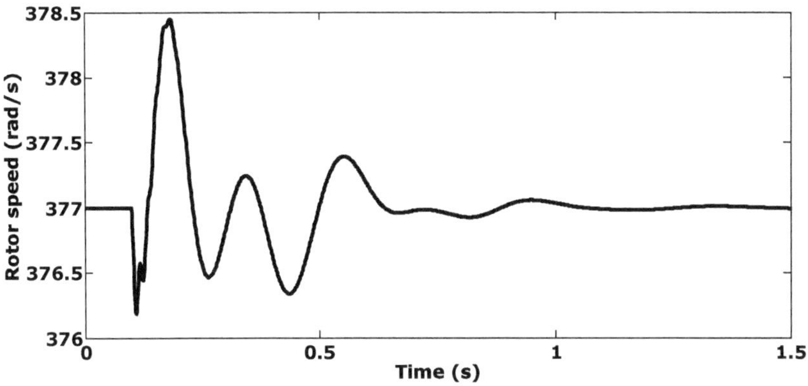

(a) Rotor speed during unknown disturbances

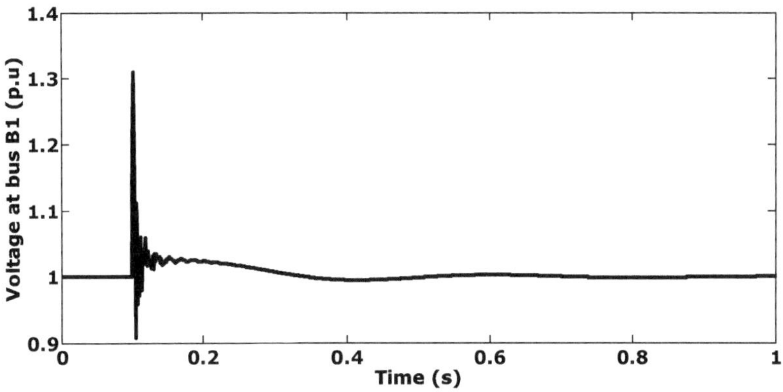

(b) Bus B1 voltage during unknown disturbances

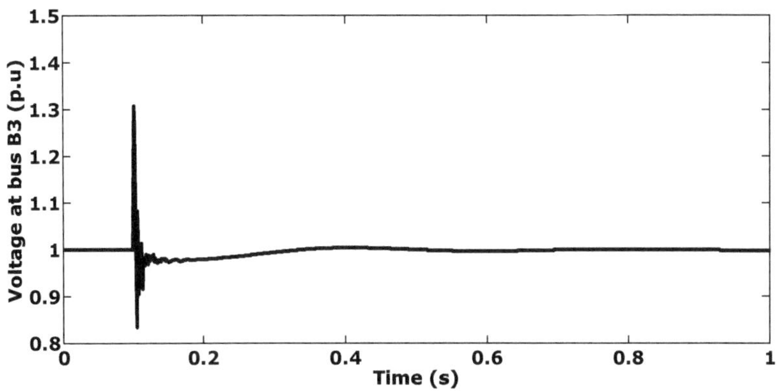

(c) Bus B3 voltage during unknown disturbances

Fig. 8.4 Performance of the robust LTI-MPC for unknown disturbances: **a** rotor speed, **b** bus B1 voltage, and **c** bus B3 voltage

References

Bemporad A, Morari M (1999) Robust model predictive control: a survey. In: Merabet A, Labib L, Ghias AM (eds) Robustness in identification and control. Springer, London, pp 207–226

Chisci L, Rossiter JA, Zappa G (2001) Systems with persistent disturbances: predictive control with restricted constraints. Automatica 37(7):1019–1028

Maestre JM, Negenborn RR (eds) (2014) Distributed model predictive control made easy, vol 69. Springer, Dordrecht, Netherlands

Mayne DQ, Seron MM, Raković SV (2005) Robust model predictive control of constrained linear systems with bounded disturbances. Automatica 41(2):219–224

Merabet A, Labib L, Ghias AM (2017) Robust model predictive control for photovoltaic inverter system with grid fault ride-through capability. IEEE Trans Smart Grid 9(6):5699–5709

Tavakoli A, Negnevitsky M, Muttaqi KM (2016) A decentralized model predictive control for operation of multiple distributed generators in an islanded mode. IEEE Trans Ind Appl 53(2):1466–1475

Tibol, JR, Pinheiro H, de Camargo RF (2011) Closed loop selective harmonic elimination applied to a grid connected PWM converter with LCL filter. In: XI Brazilian power electronics conference. IEEE, pp 746–752

Wang L (2009) Model predictive control system design and implementation using MATLAB®. Springer Science & Business Media

Yang F, Sun Q, Han QL, Wu Z (2016) Cooperative model predictive control for distributed photovoltaic power generation systems. IEEE J Emerg Select Top Power Electron 4(2):414–420

Chapter 9
Conclusion and Future Scope

Abstract The foundation of this book fundamentally lies in the fact that in a hierarchical control of the micro-grid, an advanced controller called MPC is superior in performance compared to the conventional micro-gird controllers. The superiority of the MPC at the secondary control level was already a thoroughly discussed topic in the micro-grid literature. But its application at the primary control level of a micro-grid is still a point of discussion. The MPC at the micro-grid primary level is decentralized in nature and restricted only to the linear models of the micro-grid constituents. The major implementation issue with the MPC at the primary control level of a micro-grid arises from the nonlinearity of the micro-grid model and the necessity for a centralized MPC architecture that facilitates the use of large prediction and control horizons. This book focused on the MPC formulations for centralized primary control of the micro-grids with nonlinear models.

Keywords Centralized controller · Computational complexity · Load disturbance · LTI-MPC · LTV-MPC · Nonlinear dynamics · Source intermittency

9.1 Summary of the Book

The foundation of this book fundamentally lies in the fact that in a hierarchical control of the micro-grid, an advanced controller called MPC is superior in performance compared to the conventional micro-gird controllers. The superiority of the MPC at the secondary control level was already a thoroughly discussed topic in the micro-grid literature. But its application at the primary control level of a micro-grid is still a point of discussion. The MPC at the micro-grid primary level is decentralized in nature and restricted only to the linear models of the micro-grid constituents. The major implementation issue with the MPC at the primary control level of a micro-grid arises from the nonlinearity of the micro-grid model and the necessity for a

Supplementary Information The online version contains supplementary material available at https://doi.org/10.1007/978-981-19-5852-6_9.

P. Vidyasagar and K. Shanti Swarup, *Design and Development of Model Predictive Primary Control of Micro Grids*, Springer Tracts in Electrical and Electronics Engineering, https://doi.org/10.1007/978-981-19-5852-6_9

centralized MPC architecture that facilitates the use of large prediction and control horizons. This book focused on the MPC formulations for centralized primary control of the micro-grids with nonlinear models.

In Chap. 3, a micro-grid whose mathematical model is highly nonlinear and consists of all the basic constituents of a modern-day micro-grid is considered for the analysis. A detailed mathematical model of the micro-grid was explained in Chap. 3, which can be extended to micro-grids covering large geographical areas and consists of a large number of constituents (lines, loads, and DGs). The chapter covered different reference frames used in the micro-grid model (local and global), nonlinear state-space formulation of the micro-grid model, and identified the micro-grid inputs and outputs for the formulation of the MPC.

Chapter 4 gives a detailed introduction to the MPC to the readers. The chapter discussed the linear and nonlinear formulations of the MPC. A centralized LTI-MPC was discussed in Chap. 5 for the primary control of micro-grids. At each sample, a linearized model around the operating point corresponding to that sample is used for the prediction of the micro-grid response within the prediction horizon. Due to the LTI model of the micro-grid, the original nonlinear non-convex optimal control problem of the MPC was approximated to a quadratic programming problem of polynomial time complexity. The detailed performance analysis of the LTI-MPC was presented for a single R-L load disturbance, PV source intermittency, and at a time, multiple R-L loads and induction motor load disturbances. Simulation results show that the performance of the LTI-MPC was satisfactory for all the disturbance scenarios. The LTI-MPC was robust towards the micro-grid operating points. Despite using the linearized model within the prediction horizon, the LTI-MPC handled the nonlinear dynamics and constraints in the micro-grid well.

A centralized LTV-MPC for the primary control of standalone micro-grids was discussed in Chap. 6. At each sample, within the prediction horizon, LTV-MPC linearizes the nonlinear micro-grid model around the state and input reference trajectories resulting in a linear time-variant (LTV) model. The LTV model is used for predicting the forced response of the micro-grid. The natural response is predicted by solving the nonlinear model along the state and input reference trajectories. An optimal control problem for the LTV-MPC is formulated using the complete predicted response, which is a quadratic programming problem instead of a non-convex nonlinear programming problem. The performance of the LTV-MPC was presented for a single R-L load disturbance, PV source intermittency, and at a time, multiple R-L and induction motor load disturbances. The simulation results show that the LTV-MPC was superior in performance compared to the LTI-MPC. The LTV-MPC was successful as a centralized controller in performing the load-sharing between different generators both in the steady-state and the transient period. During the disturbances, it was able to maintain the system frequency and bus voltages within limits and was successful in restoring them to their nominal values at the steady-state. It performed the complex computational tasks involved in the conventional nonlinear MPC in a simple manner using an LTV approximation model of the micro-grid at every sampling instant. However, compared to the LTI-MPC, the LTV-MPC is computationally more complex.

The application of LTI-MPC and LTV-MPC for the primary control of the micro-grids with nonlinear models reduces the computational burden involved in the nonlinear MPC. In Chap. 7, the book discussed the application of orthonormal special functions in the MPC design for further reduction of computational burden on the online MPC. The chapter discusses the approximation of the original optimal control trajectories within the control horizon with orthonormal special function networks. Two kinds of special functions are employed in the book, namely Laguerre functions and two-parameter Kautz functions. The choice of the functions depends on the nature of the control trajectories that are to be approximated. Simulation results show that the use of orthonormal special functions in the LTI-MPC and LTV-MPC designs has drastically reduced the number of optimizing variables in the optimal control problem without compromising the performance of these MPC designs. This reduction in the number of optimizing variables has improved the computational complexity of the respective MPC designs.

Apart from the mathematical formulations, the online implementation of the MPC formulations requires some supportive actions. Chap. 8 solely focused on these supportive requirements. The chapter explained the scalability of the MPC formulations and concluded that the formulations could be extended to large-scale micro-grids with a cooperative MPC structure. The selective harmonic eliminator (SHE) and state estimator can be designed independently of the MPC formulations. The chapter discussed how to choose a particular MPC formulation. The choice depends on a trade-off between computational complexity and performance. Simulation results show that the LTI-MPC with special functions is the least complex, whereas LTV-MPC with pulse operators has superior performance. Finally, the chapter discussed the integration of the disturbance compensator with the MPC formulations so that the overall MPC becomes robust to parametric uncertainties, state estimation errors, and unknown disturbances.

In short, the major takeaways from the book can be described as follows:

1. The LTI-MPC efficiently handled the nonlinearity and operational constraints in the micro-grid. However, LTI-MPC is limited to small prediction horizons and involves non-utilization of the optimal information calculated at each sample.
2. An extended "tail"-based LTV-MPC can overcome the drawbacks of the LTI-MPC.
3. The application of the orthonormal basis special functions has improved the computational viability of the LTI-MPC and LTV-MPC.
4. It can be concluded that the application of special functions in the MPC formulations has reduced the number of optimizing variables in the MPC optimal control problem without compromising the performance of the controller.

9.2 Novel Concepts in the Book

The contribution of the book to the existing literature on primary control of the micro-grids can be described in two points:

Concept 1: The LTV-MPC with extended "tail" discussed in Chap. 6 is not discussed anywhere in the micro-grid control literature. This concept helps in more accurate linear approximations of the nonlinear models of the micro-grids. This leads to better prediction of the micro-grid behaviour within the prediction horizon and hence leads to relatively better performance compared to the LTI-MPC. Also, the extended "tail" helps in wisely choosing the reference trajectories around which the linearization is performed. Since the information in the "tail" is also optimal in some sense, we can say that the information in the calculated optimal control trajectories at each sample was fully utilized.

Concept 2: The application of special functions in the formulation of the LTI-MPC and LTV-MPC was a new addition to the micro-grid control literature. This particular concept was highly beneficial for the online implementation of the LTI-MPC and LTV-MPC as centralized primary controllers in the standalone micro-grid. With a good choice of the pole locations, the number of special functions to approximate each input control trajectory within the control horizon can be drastically reduced. This certainly will reduce the number of optimal variables in the MPC control problem. The special functions also lead to simplifications in the calculations required for the formulation of the MPC control problem. The computational complexity can be greatly improved with the application of special functions, which earlier was a barrier to the online implementation of the MPC as a centralized controller for the primary control of micro-grids.

9.3 Limitations

- The centralized MPC formulations discussed in the book are limited to small-scale and medium-scale micro-grids. The scale here represents the geographical areas the micro-grids cover and the number of buses and generators involved in them. Certainly, for large-scale micro-grids, the discussed MPC formulations are difficult to implement online at the primary control level.
- The book certainly gave more emphasis on the quality of the micro-grid control. Hence the MPC formulations discussed are good for those applications where cost is not a barrier, but the quality is. But when it comes to the cost, the MPC formulations discussed in the book require high initial cost in the form of fast processors with high memory requirements and good communication infrastructure.
- The solutions of the MPC formulations are obviously from the predicted behaviour based on the linearly approximated micro-grid models. Hence the solutions are suboptimal when compared to the nonlinear MPC.

9.4 Future Scope

- Identifying the class of micro-grids where the MPC formulations discussed in the book can be implemented. The class of micro-grids where performance has much higher importance than the cost can adopt these kinds of MPC formulations.
- The study of possible hardware setups that are compatible with the theoretical formulations discussed in the book.
- Identifying the cost-reducing prospects in real-time implementation.

Appendix 1

Micro-grid Information

Micro-grid Information:

SG-DG:

Parameters in p.u:

$X_{d1} = 2.86$, $X_{q1} = 2$, $R_s = 0.0052$, $X_{\mathrm{ls}} = 0.2$, $R_D = 0.08$, $X'_{d1} = 0.7$, $X'_{q1} = 0.85$, $X''_{d1} = 0.22$, $X''_{q1} = 0.21$

Time constants of the SG-DG (s):

$T'_{do} = 3.4$, $T'_{qo} = 3.4$, $T''_{do} = 0.01$, $T''_{qo} = 0.05$, $T_E = 0.36$, $T_A = 0.1$, $T_{\mathrm{CH}} = 0.2$, $T_{\mathrm{sv}} = 0.1$, $H = 2.9$

Gains of the SG-DG:

$K_E = 0.01$, $K_A = 20$

PV-DG:

Parameters in p.u:

$R_f = 0.02$, $X_f = 0.3$, $X_{Cf} = 1.3928$, $R_t = 0.016$, $X_t = 0.115$, $R_{D3} = 0.04$

Other relevant information:

$K_p = 2$, $K_i = 20$, $C_{\mathrm{dc}} = 0.1$ H, $G = 800$ w/m^2, $T = 25$ °C, $\omega_b = 377$ rad/s

Network parameters (p.u):

$R_{14} = 0.0199$, $r_{16} = 0.0168$, $R_{15} = 0.0416$, $R_{42} = 0.03$, $R_{34} = 0.0303$, $R_{37} = 0.0408$, $R_{38} = 0.034$, $X_{14} = 0.02$, $X_{16} = 0.1155$, $X_{15} = 0.2882$, $X_{42} = 0.0873$, $X_{34} = 0.050$, $X_{37} = 0.2417$, $X_{38} = 0.1923$, $X_{C1} = 19.7239$, $X_{C2} = 19.7628$, $X_{C3} = 19.6464$, X_{C4}

$= 19.9203$, $X_{C5} = 25$, $X_{C6} = 20$, $X_{C7} = 7.6923$, $X_{C8} = 9.0909$, $R_{49} = 0.0033$, $X_{49} = 0.0266$, $R_{910} = 0.0007$, $X_{910} = 0.00148$

State-space information (p.u):

States (n) = 64/70 (islanded/grid-connected), Inputs (n_{ip}) = 4, Outputs (n_{op}) = 4

Load Information (Islanded Mode)

Minimum load parameters (p.u):

$T_{\text{mm6}} = 0.1440$, $R_{l2} = 6.5920$, $X_{l2} = 3.2960$, $R_{l5} = 6.0941$, $X_{l5} = 0.0849$, $R_{l7} = 4$, $X_{l7} = 4$, $R_{l8} = 4$, $X_{l8} = 4$, SG-DG output: $0.6248 + j0.0011$, PV-DG output: $0.0626 + j0.1095$

Maximum load parameters (p.u):

$T_{\text{mm6}} = 0.1940$, $R_{l2} = 6.5920$, $X_{l2} = 3.2960$, $R_{l5} = 4.5388$, $X_{l5} = 0.0640$, $R_{l7} = 2.7948$, $X_{l7} = 2.8276$, $R_{l8} = 2.7948$, $X_{l8} = 2.8277$, SG-DG output: $0.8299 - j0.0476$, PV-DG output: $0.0626 + j0.2867$

Load flow results (p.u):

Base MVA = $\text{MVA}_{\text{b}} = 5$, Base kV = $\text{KV}_{\text{b}} = 13.8$ on B1, SG-DG output = $0.4767 + j0.0942$, PV-DG output = $0.2154 + j0.0235$, PV array output = 0.2265, $V_1 = 1$, $V_2 = 0.994$, $V_3 = 1$, $V_4 = 0.999$, $V_5 = 1.0029$, $V_6 = 0.9673$, $V_7 = 0.9942$, $V_8 = 0.9912$, $\theta_1 = 0°$, $\theta_2 = -0.9213°$, $\theta_3 = -0.5331°$, $\theta_4 = -0.3340°$, $\theta_5 = -2.8092°$, $\theta_6 = -0.7478°$, $\theta_7 = -2.3088°$, $\theta_8 = -1.9092°$

Load parameters at the initial operating point (p.u):

$R_{\text{m6}} = 0.031$, $X'_{\text{m6}} = 0.2704$, $X_{\text{m6}} = 3.3$, $S_{\text{m6}} = 0.0028$, $T_{\text{mm6}} = 0.1440$, $R_{l2} = 6.5920$, $X_{l2} = 3.2960$, $R_{l5} = 6.0941$, $X_{l5} = 0.0849$, $R_{l7} = 3.8306$, $X_{l7} = 3.8757$, $R_{l8} = 3.8079$, $X_{l8} = 3.8527$

Time constants of the load (s):

$T_{\text{m6}} = 0.4981$, $H_{\text{m6}} = 0.7$

Controller Information (Islanded Mode)

Controller information:

Sampling time (T_s) = 1 ms, Reference set-point for the rotor speed (ω_r) = 377 rad/s (60 Hz), Reference set-point for the bus B1 voltage (V_1) = 1 p.u, Reference set-point for the bus B3 voltage (V_3) = 1 p.u, Nominal output % of the PV-DG ($P_{3,\text{nom}}$) = 85% of the PV array's maximum output, Primary reserve margin of the PV-DG = 15%,

Prediction horizon length (N_p) = 150, Control horizon length (N_c) = 150, Upper limit for the prediction horizon length from LTI analysis = 164, Upper limit for the prediction horizon length from LTV analysis = 175, Input penalty on each SG-DG input = 0.0000001, Input penalty on each PV-DG input = 0.0001, Disturbance compensator gain for each state in the compensator gain matrix $\boldsymbol{\alpha} = 500$.

Special functions information:

Number of Laguerre functions (N_s) = 50, Laguerre network pole $p = 0.5$, Number of Kautz functions (N_s) = 50, Kautz network parameters: $b = 0.7984$, $c = -0.2525$.

Base case information for calculating computational complexity:

Prediction horizon length $N_p = 100$, Control horizon length $N_c = 100$, Number of SG-DG's $N_{\text{sg}} = 1$, Number of PV-DG's $N_{\text{pv}} = 1$, Penalty on each SG-DG input = 0.0000001, Penalty on each PV-DG input = 0.0001, Sampling time $T_s = 1$ ms, Laguerre network parameters: $N_s = 50$, $p = 0.3$, Kautz network parameters: $N_s = 50$, $b = 0.5492$, $c = -0.0925$.

Appendix 2

Steps to Execute the Examples in MATLAB

Example 5.1:

Step 1: Download the folder "Example 5.1".
Step 2: Unzip the folder. The folder contains the saved workspace for the initial operating point (initialoppoint.mat).
Step 3: Open the MATLAB file named "r717LTI.m" and execute the file.
Step 4: Make sure while running the file that all the other files, namely "objective.m", "objective1.m", "objective2.m", and "initialoppoint.mat" are in the same folder as that of the "r717LTI.m" for the successful execution.
Step 5: As pointed out in Chap. 5, the file simulates the performance of the LTI-MPC for an R-L load disturbance at bus B7 of the micro-grid shown in Fig. 3.1.
Step 6: To change the disturbance level, there is a heading named "% Application of load disturbance" in the code, below which the values of Rload7 and Lload7 can be changed.
Step 7: To change the bus *i* at which the disturbance occurs, change the names of the load Rload*i* and Lload*i* (*i* is the bus number) accordingly under the heading "% Application of load disturbance".
Step 8: Once the execution of "r717LTI.m" is completed, the results for different micro-grid outputs are stored in the following variables:

Rotor speed → y1new

Bus1 voltage → y2new
Bus3 voltage → y3new
SG-DG output → y4new
PV-DG output → y5new

Step 9: To plot any of the above outputs, type the command "plot(knew,output variable name)" in the command window of the MATLAB. For example, to get the plot of the rotor speed, type the command "plot(knew,y1new)".

Example 5.2:

Step 1: Download the folder "Example 5.2".
Step 2: Unzip the folder. The folder contains the saved workspace for the initial operating point (initialoppoint.mat).
Step 3: Open the MATLAB file named "im6r5l5r7l7r8l8LTI.m" and execute the file.
Step 4: Make sure while running the file that all the other files, namely "objective.m", "objective1.m", "objective2.m", and "initialoppoint.mat" are in the same folder.
Step 5: As pointed out in Chap. 5, the file simulates the performance of the LTI-MPC for multiple R-L and IM load disturbances in the micro-grid shown in Fig. 3.1.
Step 6 to Step 9 are similar to Example 5.1.

Example 5.3:

Step 1: Download the folder "Example 5.3".
Step 2: Unzip the folder. The folder contains the saved workspace for the initial operating point (initialoppoint.mat).
Step 3: Open the MATLAB file named "intermittencyLTI.m" and execute the file.
Step 4: Make sure while running the file that all the other files, namely "objective.m", "objective1.m", "objective2.m", and "initialoppoint.mat" are in the same folder.
Step 5: As pointed out in Chap. 5, the file simulates the performance of the LTI-MPC for the PV-DG source intermittency in the micro-grid shown in Fig. 3.1.
Step 6: To change the intermittency level, there is a heading named "% Application of source intermittency as input disturbance" in the code, below which the value of the irradiance G1 can be changed.
Step 7: Continue Step 8 and Step 9 of Example 5.1.

Example 6.1:

Step 1: Download the folder "Example 6.1".
Step 2: Unzip the folder. The folder contains the saved workspace for the initial operating point (initialoppoint.mat).
Step 3: Open the MATLAB file named "r7l7LTVnatural.m" and execute the file.

Step 4: Make sure while running the file that all the other files, namely "objective.m", "objective1.m", "objective2.m", and "initialoppoint.mat" are in the same folder.
Step 5: As pointed out in Chap. 6, the file simulates the performance of the LTV-MPC for an R-L load disturbance at bus B7 of the micro-grid shown in Fig. 3.1.
Step 6 to Step 9 are similar to Example 5.1.

Example 6.2:

Step 1: Download the folder "Example 6.2".
Step 2: Unzip the folder. The folder contains the saved workspace for the initial operating point (initialoppoint.mat).
Step 3: Open the MATLAB file named "im6r5l5r7l7r8l8LTVnatural.m" and execute the file.
Step 4: Make sure while running the file that all the other files, namely "objective.m", "objective1.m", "objective2.m", and "initialoppoint.mat" are in the same folder.
Step 5: As pointed out in Chap. 6, the file simulates the performance of the LTV-MPC for multiple R-L and IM load disturbances in the micro-grid shown in Fig. 3.1.
Step 6 to Step 9 are similar to Example 5.1.

Example 6.3:

Step 1: Download the folder "Example 6.3".
Step 2: Unzip the folder. The folder contains the saved workspace for the initial operating point (initialoppoint.mat).
Step 3: Open the MATLAB file named "intermittencyLTVnatural.m" and execute the file.
Step 4: Make sure while running the file that all the other files, namely "objective.m", "objective1.m", "objective2.m", and "initialoppoint.mat" are in the same folder.
Step 5: As pointed out in Chap. 6, the file simulates the performance of the LTV-MPC for the PV-DG source intermittency in the micro-grid shown in Fig. 3.1.
Step 6: To change the intermittency level, there is a heading named "% Application of source intermittency as input disturbance" in the code, below which the value of the irradiance G1 can be changed.
Step 7: Continue Step 8 and Step 9 of Example 5.1.

Example 7.1:

Step 1: Download the folder "Example 7.1".
Step 2: Unzip the folder. The folder contains the saved workspace for the initial operating point (initialoppoint.mat).
Step 3: The folder contains 9 files that need to be executed, out of which 3 files are for the pulse operators (r7l7LTI.m, im6r5l5r7l7r8l8LTI.m, intermittencyLTI.m), 3 for the Laguerre operators (r7l7LTIlag.m, im6r5l5r7l7r8l8LTIlag.m,

intermittencyLTIlag.m) and 3 for the Kautz operators (r7l7LTIkautz.m, im6r5l5r7l7r8l8LTIkautz.m, intermittencyLTIkautz.m).
Step 4: Execute the 9 files one after the other.
Step 5: Make sure while running the files that all the other files, namely "objective.m", "objective1.m", "objective2.m", "kautzd.m", "lagd.m", and "initialoppoint.mat" are in the same folder.
Step 6: As pointed out in Chap. 7, the files simulate the performance of the LTI-MPC with the special functions for the different load disturbance and source intermittency scenarios in the micro-grid shown in Fig. 3.1.
Step 7: Continue Step 8 and Step 9 of Example 5.1.

Example 7.2:

Step 1: Download the folder "Example 7.2".
Step 2: Unzip the folder. The folder contains the saved workspace for the initial operating point (initialoppoint.mat).
Step 3: The folder contains 9 files that need to be executed, out of which 3 files are for the pulse operators (r7l7LTVnatural.m, im6r5l5r7l7r8l8LTVnatural.m, intermittencyLTVnatural.m), 3 for the Laguerre operators (r7l7LTVnaturallag.m, im6r5l5r7l7r8l8LTVnaturallag.m, intermittencyLTVnaturallag.m) and 3 for the Kautz operators (r7l7LTVnaturalkautz.m, im6r5l5r7l7r8l8LTVnaturalkautz.m, intermittencyLTVnatural.m).
Step 4: Execute the 9 files one after the other.
Step 5: Make sure while running the files that all the other files, namely "objective.m", "objective1.m", "objective2.m", "kautzd.m", "lagd.m", and "initialoppoint.mat" are in the same folder.
Step 6: As pointed out in Chap. 7, the files simulate the performance of the LTV-MPC with the special functions, for the different load disturbance and source intermittency scenarios in the micro-grid shown in Fig. 3.1.
Step 7: Continue Step 8 and Step 9 of Example 5.1.

MIX
Papier aus verantwortungsvollen Quellen
Paper from responsible sources
FSC® C105338

If you have any concerns about our products,
you can contact us on
ProductSafety@springernature.com

In case Publisher is established outside the EU,
the EU authorized representative is:
Springer Nature Customer Service Center GmbH
Europaplatz 3, 69115 Heidelberg, Germany

Printed by Libri Plureos GmbH
in Hamburg, Germany